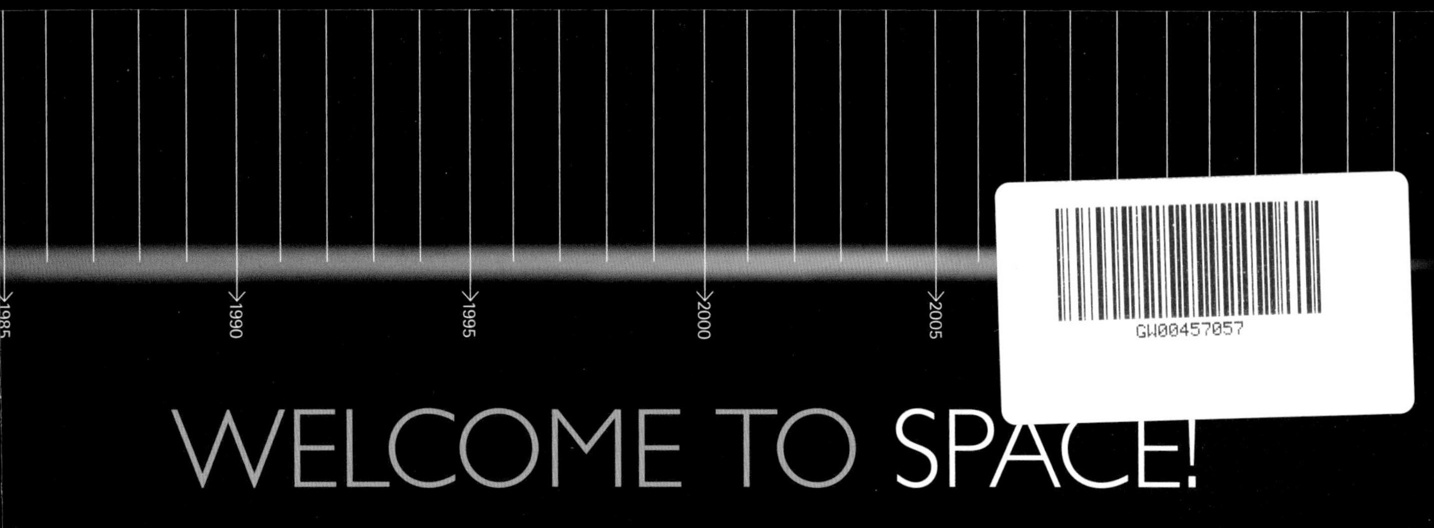

WELCOME TO SPACE!

It's just 60 short years since the first man to reach space soared high above the flat dustbowl of Kazakhstan. The date was April 12, 1961 and the man's name was Yuri Gagarin. Within a month an American too had rocketed into space. But it was a pitiful 15-minute flight – straight up and straight down, in a capsule one-quarter the size of the Russian Vostok. However, President Kennedy was frustrated with Soviet successes, and within two weeks of that short, sub-orbital flight he announced to the world that he was challenging the Russians to a race for the moon.

The next decade would light up the space race as both sides surged forward, each seeking to be the 'first' and to be the best. Along the way disasters took the lives of American astronauts and Russian cosmonauts, but the pace quickened in the race for the moon. Then in July 1969, Neil Armstrong took his 'one small step' and transformed it into a 'giant leap for mankind' as he put the first human boot print in the lunar dust.

This is the story of that adventure, above all a story of great accomplishments, of giant telescopes launched by the Shuttle bringing views of the most distant parts of our galaxy, of medical experiments on astronauts that helps combat the ageing process in earthlings, and of routine space operations with a crew of six as the permanent occupants of a 400 ton space station.

It's all here - the intoxication of success, the tears over fallen comrades and the hopes of a generation whose aspirations soared to the stars. It is on their shoulders that a new generation will pick up the baton and carry the race forward to fresh goals and new destinations, in mankind's never ending search for new destinations in space.

David Baker

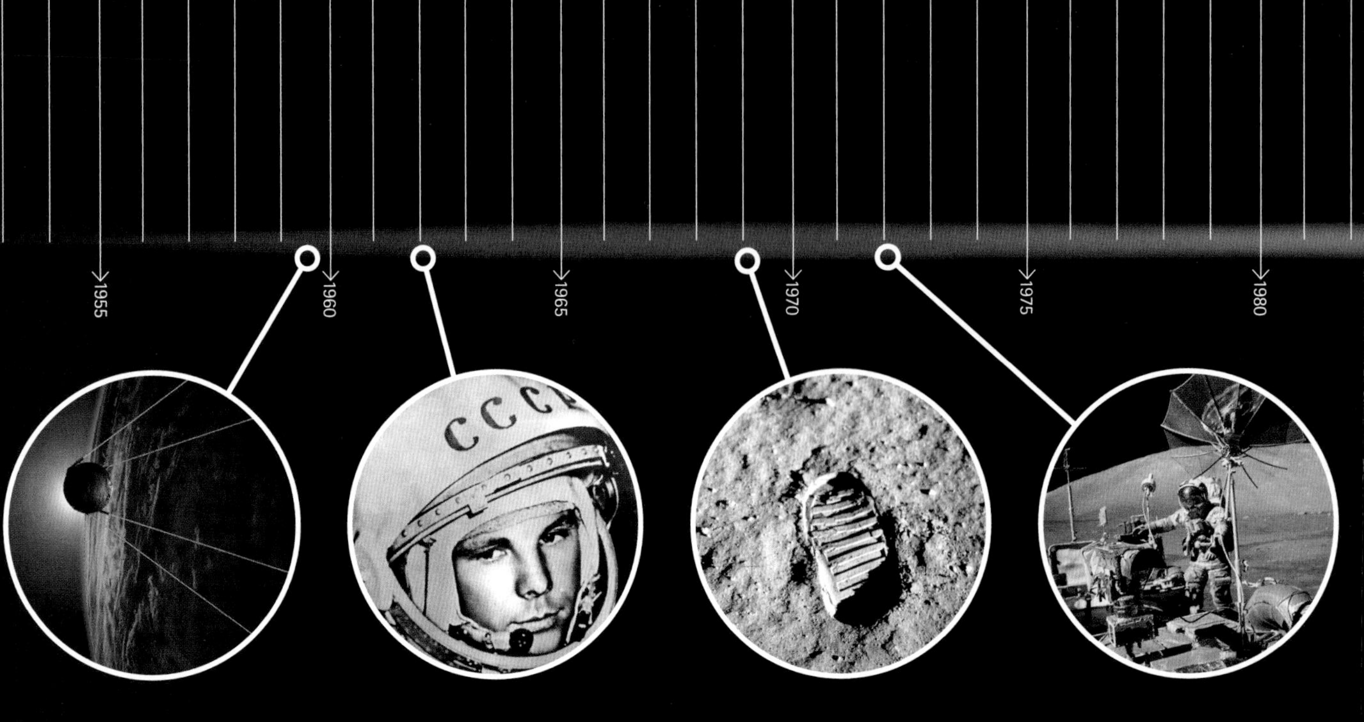

6 ROCKET FLIGHT
The road to space is paved with experimental rocket flights after World War 2.

10 A NEW DAWN FOR MANKIND
Russia gets the jump on America with the launch of Sputnik, and again with a dog called Laika.

14 NASA'S CHALLENGE
The newly formed US space agency sets its sights on putting the first man in space.

16 RED STAR IN SPACE
Again the Russians trounce the US when Yuri Gagarin heads for a full orbit of the earth.

18 MERCURY FLIES
NASA finally gets its Mercury astronauts into orbit – late and behind the Russians.

26 TO THE MOON
President Kennedy sets a moon landing as the new goal and a national effort swings into action.

28 SOVIETS SURGE INTO SPACE
Cosmonauts fly dual missions to orbit and Russia launches its first spacewoman.

30 FLOATING ON SPACE
Leonov becomes the first man to walk in space but the Russians are slowly losing the race.

32 TWINS IN ORBIT
Ten two-man Gemini missions sends NASA soaring into the lead as astronauts practice moon techniques.

44 FATEFUL YEAR
Momentum is broken when three US astronauts die in a launch pad fire, then a Russian dies on a space mission.

46 APOLLO GETS READY
Test missions start preparing the rockets and the spacecraft for a moon landing before the end of the decade.

48 MOONWALK USA
Man on the moon! America achieves its goal of beating the Russians to the lunar surface.

60 RENDEZVOUS WITH A ROBOT
On the second moon landing NASA sets down Apollo 12 alongside a previously launched robot.

Writer and editor: *David Baker*
Senior editor, Bookazines: *Roger Mortimer*
Email: roger.mortimer@keypublishing.com
All photographs and illustrations supplied by David Baker
Revised second edition. Originally published in 2011 as *SPACE - Celebrating 50 Years of Human Space Flight*

Head of design: *Steve Donovan*
Design: *Andy O'Neil, Dave Robinson, Debbie Walker*
Head of production: *Janet Watkins*
Advertising group manager: *Brodie Baxter*
Advertising production manager: *Debi McGowan*
Marketing manager: *Shaun Binnington*

Group CEO: *Adrian Cox*
Publisher: *Mark Elliott*
Head of publishing: *Finbarr O'Reilly*
Chief publishing officer: *Jonathan Jackson*

Key Publishing Ltd
PO Box 100, Stamford, Lincs, PE9 1XQ
Tel: 01780 755131 Fax: 01780 757261
Email: enquiries@keypublishing.com
www.keypublishing.com

Distribution: *Seymour Distribution Ltd, 2 Poultry Avenue, London EC1A 9PP. Tel: 020 7429 4000*
Printed by: *PCP Ltd, Haldane, Halesfield 1, Telford. TF7 4QQ*

Published by: Key Publishing Ltd • PRINTED IN ENGLAND

CONTENTS

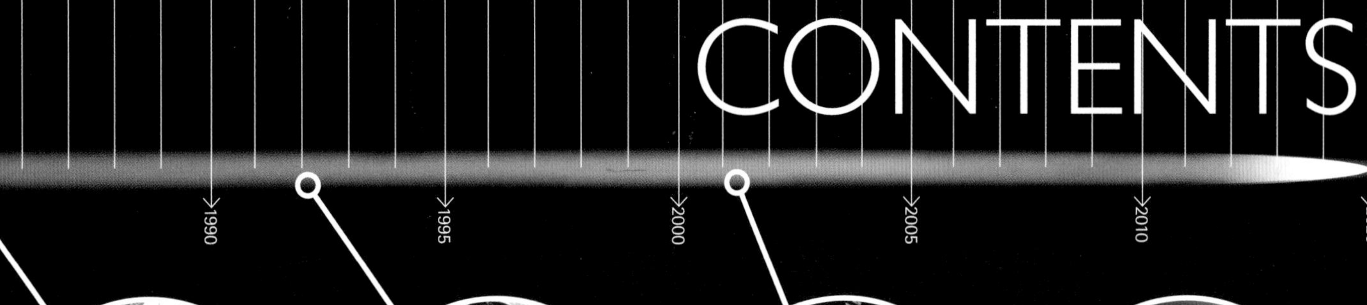

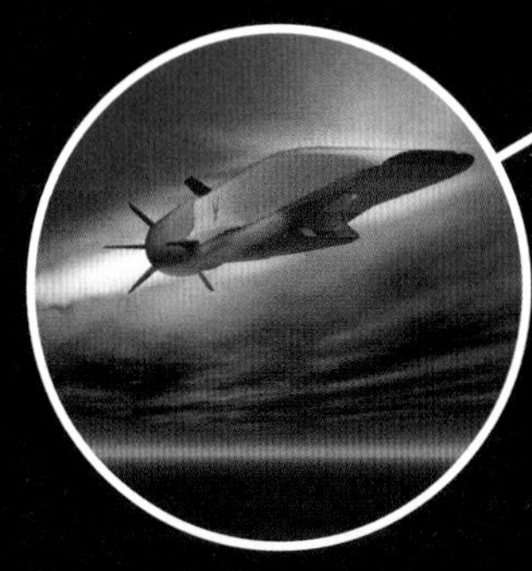

62 HOUSTON – WE HAVE A PROBLEM
Heading toward the moon, Apollo 13 runs into trouble and limps home with an exhausted crew.

64 HEADING FOR THE HILLS
Complete with handcart, the first American in space commands a mission to the moon's Fra Mauro hills.

66 THE ELECTRIC MOON BUGGY
Three ambitious moon missions take roving vehicles and roam new and exciting sites for science.

72 THE MOON UNVEILED
Just what did we get back from the moon and what did it tell us? What do we know of its origin?

74 CONSPIRACY THEORIES
Did NASA astronauts really land on the moon, or was it all mocked up in a studio? Where is the evidence?

76 ROOMS WITH A VIEW
Left-over Apollo hardware gives astronauts the chance to stay in space for up to three months.

86 EAST MEETS WEST
Détente in space as the Americans and Russians dock in orbit.

90 WINGS INTO SPACE
Switching to the reusable Shuttle, NASA prepares for a new age of winged space flight.

94 GENESIS OF A NEW SPACE AGE
Shuttle trucks carry satellites and payloads into space and a European laboratory for scientists.

109 TRAGEDY STRIKES
A faulty booster destroys Challenger shortly after launch, killing all seven on board.

112 AN EYE ON THE UNIVERSE
Restored to flight, the Shuttle launches the biggest orbiting telescope and keeps it updated for a decade.

116 MIR- RUSSIA'S HOME IN SPACE
Cosmonauts assemble Russia's biggest space station and stay in orbit for long periods equaling a trip to Mars.

120 A GLOBAL VIEW FROM SPACE
Former adversaries joined forces and lead international partners in building the biggest laboratory in orbit.

128 FUTURESPACE
What now for the human exploration of space? Where are we going next?

130 WINDOWS ON THE PAST – VIEWS OF THE FUTURE
Places to go and websites to search for additional information.

ROCKET FLIGHT

Rocket flight becomes a reality during World War 2, pioneering the age of supersonic flight

With broken ribs from a riding accident, using a sawn-off broom handle as a tool to fasten the cockpit door his injuries prevented him from operating, 'Chuck' Yeager struggled into the world's fastest rocket ship and broke the sound barrier in level flight for the first time. He was Major Charles E Yeager, USAF - the world's first supersonic man and the date was October 14, 1947.

Almost exactly ten years later the world's first artificial satellite – Sputnik 1 – was launched into space. Between those two events pilots, engineers and scientists struggled to master high speed flight. So high speed in fact that in America they were already building the world's first spaceplane when Russia put that first satellite into space.

These were the learning years, a decade of trying to understand what it would take to reach space and fly back down again to a conventional landing. Science fiction writers had dreamed about this age for generations and they were about to get

A converted B-50 is jacked up to receive the Bell X-1 prior to an air-drop flight.

Above: X-15 test pilots, left to right: Joseph Engle, Robert Rushworth, Jack McKay, Pete Knight, Milton Thompson, Bill Dana. Right: Chuck Yeager in Glamorous Glennis, the XS-1 in which he broke the sound barrier.

Above: The father of the science of supersonic flight, Theodore von Karman in his laboratory in the United States after leaving Germany in the 1930s.

Test pilot Joseph Walker 'mounts' his Bell X-1A.

Milestones

1947
OCTOBER 14: CHUCK YEAGER IN XS-1 BREAKS THE SOUND BARRIER FOR THE FIRST TIME IN LEVEL FLIGHT.

1953
NOVEMBER 20: SCOTT CROSSFIELD IN DOUGLAS D-558-II BECOMES THE FIRST MAN THROUGH MACH 2.

1956
SEPTEMBER 27: MEL APT FIRST THROUGH MACH 3 IN BELL X-2.

1959
SEPTEMBER 17: FIRST POWERED FLIGHT OF THE X-15 ACHIEVES MACH 2.1.

1967
OCTOBER 17: FASTEST X-15 FLIGHT AT MACH 6.7 (4,520MPH)

their predictions realized. But first there was the sound barrier – then the heat barrier – and both were seemingly impregnable walls.

No longer the limit

To push aircraft through the sound barrier, about 760mph at sea level, takes a powerful force and while the jet engine was well known and tested by the end of the Second World War, at first only the rocket motor promised sufficient thrust to push an aircraft through the speed of sound. Many scientists had predicted the effects of flying faster than sound, or Mach 1, notable among which was the Hungarian Theodore von Karman who provided calculations to tackle the problem scientifically.

In March 1945 Bell Aircraft was contracted by the US Army Air Force and the National Advisory Committee for

First through Mach 3, the ill-fated Bell X-2 after a landing accident.

Aeronautics (NACA) to build a research aircraft capable of cracking the sound barrier. The result was the XS-1, powered by a rocket motor with a thrust of 6,000lb, in which Chuck Yeager became the first supersonic pilot in level flight. There were several earlier claimants to having dived through Mach 1 – but few had lived to talk about it!

Flying the heat wall

Progress was relatively slow. In the early 1950s the Bell X-2 was designed to fly at three times the speed of sound, pushing on the thermal barrier created by a build up of heat from friction with the atmosphere. By this time the NACA was thinking about extremely high altitude flight and reaching to the edge of space and on September 7, 1956, Ivan Kincheloe became the first pilot to exceed an altitude of 100,000ft. Less than three weeks later Mel Apt was killed trying to make a high speed turn but managed to achieve an unofficial record of Mach 3.2 before crashing.

Hypersonic flight

The thermal barrier was the target for the North American X-15, designed to withstand temperatures of 1,200 deg F by reaching beyond Mach 5, hypersonic flight. In fact it would reach a top speed of Mach 6.7 and a maximum height of 354,000ft, over 67 miles. Powered by a rocket motor with a thrust of 70,000lb – more than that of the German V-2 ballistic missile, in 1963 pilot Joe Walker flew this remarkable research aircraft above 100km (62.1 miles) recognized internationally as the beginning of space. The NACA and the Air Force wanted to push on with a developed version of the X-15, launched by a converted ballistic missile directly into space. But for that, with high heat incurred during re-entry from 17,500 mph, there would need to be new coatings for protecting the Inconel-X out of which the airframe was built. But before those plans could be completed, the Russians sprang a surprise – Sputnik 1 – and everything changed.

Test pilot Milton Thompson with the hypersonic North American X-15.

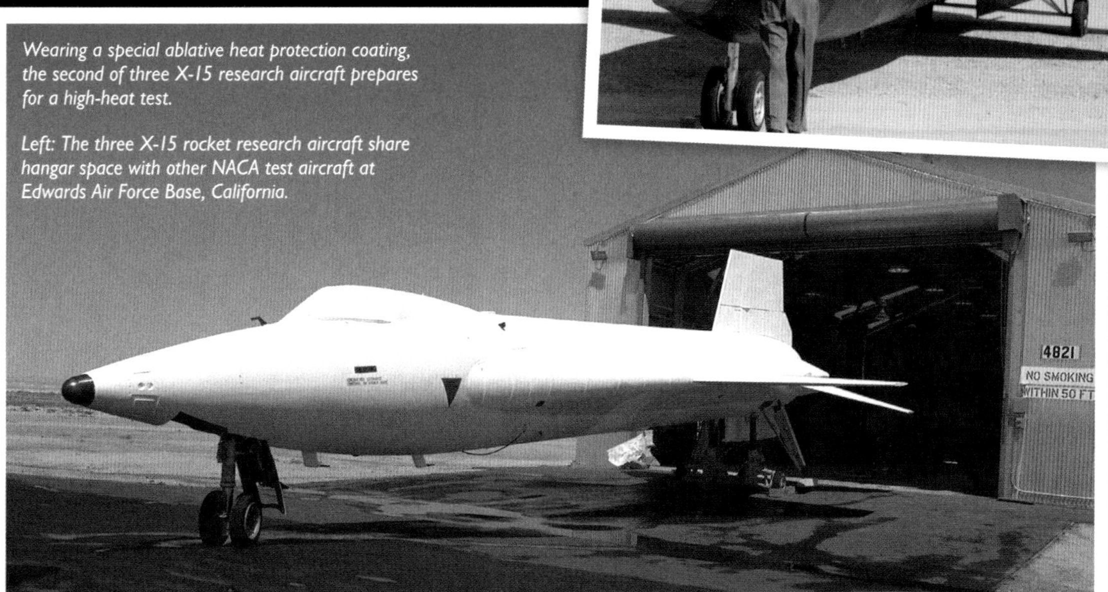

Wearing a special ablative heat protection coating, the second of three X-15 research aircraft prepares for a high-heat test.

Left: The three X-15 rocket research aircraft share hangar space with other NACA test aircraft at Edwards Air Force Base, California.

Milestones

1957

MAY 15:
FIRST LAUNCH ATTEMPT FOR RUSSIA'S R-7 BALLISTIC MISSILE ENDS IN FAILURE.

AUGUST 21:
FIRST SUCCESSFUL FLIGHT OF THE R-7.

OCTOBER 4:
SPUTNIK I IS LAUNCHED FROM A SECRET BASE IN KAZAKHSTAN.

NOVEMBER 3:
RUSSIA'S LAUNCH SPUTNIK 2 CARRYING THE DOG LAIKA.

DECEMBER 6:
FIRST ATTEMPT TO LAUNCH A US SATELLITE FAILS.

1958

JANUARY 31:
EXPLORER I LAUNCHED BY WERNHER VON BRAUN'S JUPITER-C.

LEFT • Sputnik 2 got this special stamp acknowledging the launch of the first living thing into orbit, the three-year old dog Laika collected as a stray from the streets of Moscow.

A NEW DAWN FOR HUMANKIND

Nobody took much notice of the news that America would launch a satellite - then the Russians launched one first and everything changed.

News about the launch of the world's first satellite broke on October 4, 1957, with a bleeping sound from Radio Moscow announcing that the Soviet Union had joined the ranks of the technological elite. The influence of Sputnik 1 far outweighed its importance but the fear it gave to unsuspecting Americans sparked off a space race unlike anything seen before in peacetime.

For several years Americans had been experimenting with rockets, but only as recently as 1954 had a special committee authorized a crash programme to build inter-continental ballistic missiles capable of lobbing atomic warheads to the Soviet Union. Now the Russians were demonstrating that they too were capable of building big rockets. If they could launch satellites into space they could hit the US with atomic weapons. The race was on – to build larger missiles and bigger arsenals, and to demonstrate their prowess before an awe-struck world.

The IGY

The decision to launch satellites had been taken in the early 1950s to support a worldwide study of the earth, the atmosphere and outer space using all forms of technical wizardry, including artificial satellites sending data back to the ground. The period of study was to begin on July 1, 1957 and last until the end of 1958. It was known as the International Geophysical Year, or IGY, and it would involve all the major countries around the world. Both Russia and America announced that they would launch satellites – but nobody thought the Russians could do it.

Sputnik

The push to build rockets began under Soviet leader Josef Stalin. As development of the first Russian intercontinental missile, the R-7, made progress the idea of using it to launch a satellite was the brainchild of

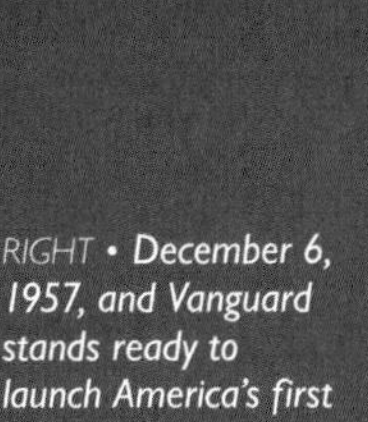

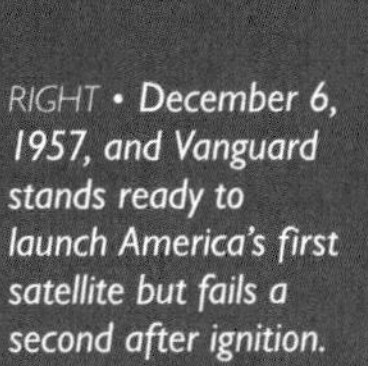

RIGHT • *A stamp issued to celebrate the launch of Sputnik 1, the world's first satellite.*

RIGHT • *December 6, 1957, and Vanguard stands ready to launch America's first satellite but fails a second after ignition.*

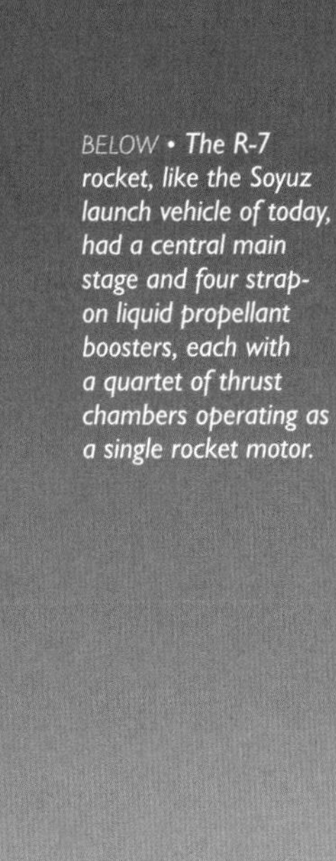

BELOW • *The R-7 rocket, like the Soyuz launch vehicle of today, had a central main stage and four strap-on liquid propellant boosters, each with a quartet of thrust chambers operating as a single rocket motor.*

Sergei Korolev, a brilliant Russian aircraft and rocket engineer. It was he who convinced the then Russian premier Nikita Krushchev to approve the satellite project, using the IGY to publicise Soviet capabilities.

The launch vehicle for Sputnik 1 was to be the R-7, easily capable of putting the 184lb satellite in orbit. Initial tests of the R-7 met with failure before the first successful flight of the rocket in August 1957, clearing the way for the satellite launch. When it had been accomplished Kruschchev telephoned Korolev to congratulate him, learning that the space group wanted to put up a dog in Sputnik 2. Gaining immediate approval, the dog Laika was launched aboard the 1,102lb Sputnik 2 in early November. It died within days.

Flopnik

American scientists had been working on their own satellite launcher, developed from the Viking research rocket and called Vanguard. The first launch attempt came n December 6 from Cape Canaveral but it fizzled out and fell over, the press hailing it as Flopnik 1! Prevented from using the Army's Jupiter C research rocket to launch a satellite because President Eisenhower wanted it to be seen as a civilian science venture, Wernher von Braun was now given the go-ahead.

On January 31, a Jupiter-C soared skyward sending Explorer 1 into orbit. Honour was restored, but it was not sufficient to assuage an angry public and politicians in Washington DC. They wanted action, and fast, not only to restore America's pride but to put the US first. The race was on.

On March 16, 1926, the American scientist Robert Goddard became the first person to launch a liquid propellant rocket to a height of 40ft and directly into a cabbage patch 184ft away!

TOP LEFT • In Britain, Jodrell Bank radio telescope in Cheshire played an important role in tracking Soviet satellites. When unmanned space probes were sent to the moon, they often downloaded the images and distributed pictures to the British press before the Russian announcements! LEFT • Launch activity at Cape Canaveral catches a thunderstorm across the gantries, converted oil derricks from the Gulf of Mexico. RIGHT • Success! From Cape Canaveral, the Army's Wernher von Braun launches the first US satellite at 10.48pm local time, January 31, 1958.

" BOTH RUSSIA AND AMERICA ANNOUNCED THAT THEY WOULD LAUNCH SATELLITES – BUT NOBODY THOUGHT THE RUSSIANS COULD DO IT. "

NASA's CHALLENGE

It was not the best way to get into orbit but it was the quickest way to beat the Russians - and that's what mattered.

RIGHT • The Mercury spacecraft was tiny, only 6ft in diameter with a weight on orbit of little more than 3,000lb and no ability to change its trajectory other than to fire retro-rockets for return to earth.

BELOW • NASA's first headquarters building was the Dolley Madison House, left by the deceased President James Madison to his wife in 1836. In 1952 it was bought by the US government for offices.

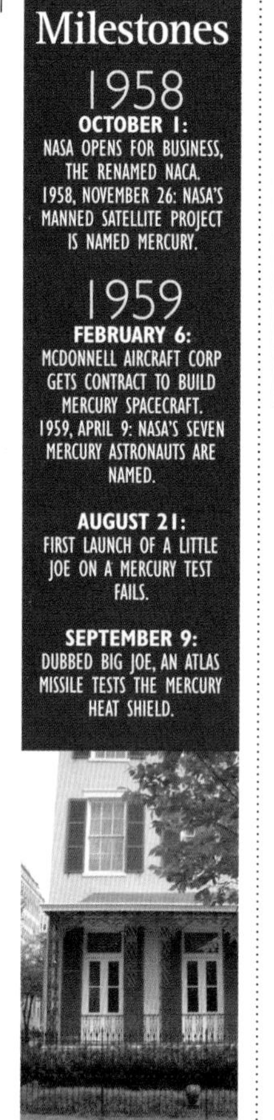

Milestones

1958

OCTOBER 1:
NASA OPENS FOR BUSINESS, THE RENAMED NACA.
1958, NOVEMBER 26: NASA'S MANNED SATELLITE PROJECT IS NAMED MERCURY.

1959

FEBRUARY 6:
MCDONNELL AIRCRAFT CORP GETS CONTRACT TO BUILD MERCURY SPACECRAFT.
1959, APRIL 9: NASA'S SEVEN MERCURY ASTRONAUTS ARE NAMED.

AUGUST 21:
FIRST LAUNCH OF A LITTLE JOE ON A MERCURY TEST FAILS.

SEPTEMBER 9:
DUBBED BIG JOE, AN ATLAS MISSILE TESTS THE MERCURY HEAT SHIELD.

Few people in the United States expected Russia to launch a satellite first and within a few weeks the US government decided it needed a special agency to look after America's scientific space programmes. For that, President Eisenhower insisted it should be a civilian agency of the government and for that Congress decreed that on October 1, 1958, the National Advisory Committee for Aeronautics, the NACA, should become the National Aeronautics and Space Administration, NASA. The NACA had been formed in 1915 to manage the technical and scientific development of aeronautics and now it would do more of the same and embrace space exploration as well.

The first job for NASA was to take hold of plans for a manned satellite, shortly thereafter named Mercury after the Roman messenger of the gods. With no certainty, it was hoped that Mercury would place the first man in space before the Russians. Time was of the essence and the fastest way to do that was to bypass the favoured route of building a winged successor to the X-15 and use existing ballistic missile technology to put a man in a can and hurl him spaceward on a converted intercontinental ballistic missile.

Testing times

The Mercury plan envisaged several sub-orbital hops where the capsule and occupant would be fired vertically to the edge of space by a smaller Redstone rocket, essentially the same vehicle as used to launch the first US satellite. Later, when the tiny spacecraft was seen to work, a much

INSERT LEFT • *Selected for their short height and long flying experience, the seven Mercury astronauts were test pilots from the three armed services. Left to right: Carpenter, Cooper, Glenn, Grissom. Schirra, Shepard, Slayton.*

Above: Fired by a Little Joe solid propellant rocket on August 21, 1959, the first operational Mercury test was this mock-up spacecraft. Set up to evaluate the launch escape system, the test was a failure.

BELOW • *Atlas was adapted to place Mercury in orbit but the missile was itself undergoing development and many were failing. This successful test launch took place on February 21, 1961 and hurled the Mercury spacecraft a distance of more than 1,000 miles across the Atlantic Ocean from Cape Canaveral.*

more powerful Atlas rocket would place an astronaut in orbit. The first test came in August 1959 when a Little Joe rocket was fired to test the capsule's escape system, a rocket held on top by a lattice tower, to wrench the capsule free of an ascending booster should anything go wrong. It flopped. The next attempt called Big Joe was for an Atlas missile to hurl a Mercury mock-up from Cape Canaveral to test the heat shield. This was a partial success and was followed by several more tests during 1960.

The Right Stuff

By now the seven Mercury astronauts had been selected from more than 500 applicants, names that would be immortalized as America's first spacemen: Scott Carpenter, John Glenn, Gordon Cooper, Virgil Grissom, Walter Schirra, Alan Shepard and Deke Slayton. Only Slayton would not make a Mercury flight, waiting 16 years to fly the joint docking mission with the Russians. Cooper and Grissom would fly Gemini also and Schirra would fly Gemini and Apollo as well. Shepard would go on to command the Apollo 14 moon landing.

From the outset, the 'magnificent 7' were given legendary status. Life magazine paid for their stories and profits from that was divided equally among them all. Five of their forenames were given to characters in the TV series Thunderbirds, and in the 1979 book of the same name, writer Tom Wolfe characterized them as being chosen for having The Right Stuff. But it would take more than that to ride rockets designed by contractors with the lowest bid, where almost one in two test launches ended in disaster.

FAR LEFT • *The NACA had been formed on March 3, 1915, to provide aircraft designers with scientific research data.* LEFT • *NASA was formally authorised to replace the NACA from October 1, 1958, in direct response to a national outrage at Russia's Sputnik launches.*

RED STAR

Sputnik 1 laid down a challenge among equals – then the Russians upped their game with the first man in orbit.

TOP LEFT: Gagarin's scorched re-entry module shows evidence of intense heat.

TOP MIDDLE: Vostok consisted of a spherical re-entry module attached to a circular equipment section which carried a retro-rocket. The large white cylindrical object is the second stage of the launch vehicle.

TOP RIGHT: The two-stage Vostok launcher was a derivative of that used to launch Sputnik 1.

RIGHT: Gagarin prepares to take the elevator to the Vostok 1 spacecraft – April 12, 1961.

The Russian design engineer Sergei Korolev was confident they had the Americans on the run, but now he wanted to put a man in orbit and to do it first. The Russian spacecraft would be called Vostok but it would also be used to carry cameras in a secret, unmanned version that on other flights would operate as a spy satellite. The module (otherwise designed to return a human from space) being used on those missions to bring back valuable film. The spy version would be known as Zenit and the first in a long line of such satellites would be launched on December 11, 1961, the last on June 7, 1994.

Getting ready

Vostok was built in two sections attached to the second stage of an adapted R-7 rocket similar to that which had been used to launch Sputnik. The 7.5ft diameter spherical re-entry module carried the cosmonaut and weighed 5,400lb with an ablative heat protection coating that would char away during re-entry. The equipment section carried a retro-rocket that would be separated prior to re-entry and had a diameter of 7.5ft in diameter and a length of just over 7ft. Together the two components weighed 10,400lb.

Seven test launches were made with Vostok modules between May 1960 and March 1961 but only three were successful. The odds were stacked against Yuri Gagarin when he walked to his rocket on the morning of April 12, 1961. He was one of 20 candidates recruited in March 1960 of which six were fast-tracked for flights. These men

ABOVE: *Born on March 9, 1934, Gagarin was drawn to flying at an early age and died in an air crash on March 27, 1968.*

BOTTOM: *The Vostok control panel was simplicity itself, providing little opportunity for manual control by the pilot.*

N SPACE

reported to the training centre, a purpose-built facility known as Zvyozdny Gorodok – Star City – just 20 miles outside Moscow. Two men stood out. Gagarin was one of them, the second being Gherman Titov.

Gagarin flies

The first man in space was launched at 9.07am local time on April 12, 1961, from the same pad at Baikonur cosmodrome used to launch Sputnik 1. Neither Gagarin nor Korolev had been able to sleep that night. Tension was almost unbearable but within five minutes Gagarin was in orbit and radioing that 'everything is working very well...everything is good...let's keep going!' Toward the end of the first orbit the guidance system swung Vostok 1 into the correct attitude for re-entry and the rocket motor fired for 42 seconds, bringing the spacecraft back down through the atmosphere and a punishing deceleration of almost 10g.

At a height of 23,000ft the hatch was jettisoned and seconds later Gagarin ejected. At 8,200ft the Vostok parachute deployed and the capsule hit the ground, bounced and rolled to a stop. Nearby two schoolgirls witnessed the historic event and a short distance away a farmer and his daughter saw a man in an orange suit and a white helmet walk toward them. As they backed away in fear, imagining him to be an alien, he opened his visor and told them 'Don't be afraid. I am a Soviet citizen like you who has descended from space – I must find a telephone and call Moscow!'

Reaction

When Gagarin returned to Moscow during the afternoon of April 12, his life had changed for ever and he would never be the same again. A world waited to applaud and fete him in countries around the globe, acclaiming this second major Soviet coup over the Western democracies with this first human from space.

In America it was still night when news of the flight broke across the air waves and a reporter called the press officer at Cape Canaveral to ask for comment. The reply he got was used in headlines across the United States and came back many times to bite NASA: 'Don't you know what time it is – we're all asleep down here!'

ССР
ВОСТОК

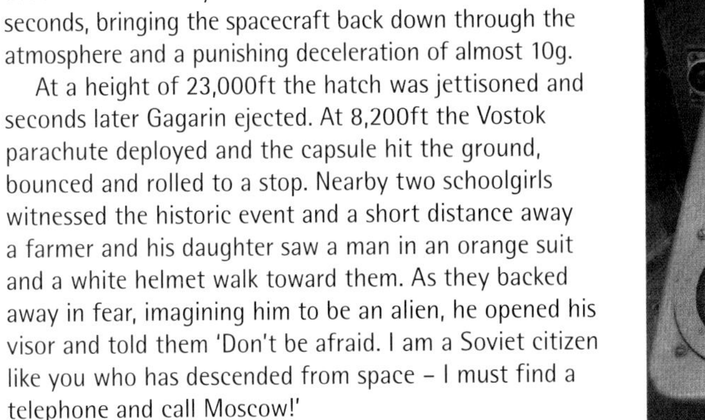

Milestones

1958

FEBRUARY 15
KOROLEV BEGINS WORK ON A MANNED SPACECRAFT TO BE CALLED VOSTOK.

SEPTEMBER 15
PRELIMINARY DESIGN OF A DUAL MANNED SPACECRAFT/ UNMANNED SPY SATELLITE COMPLETED KNOWN AS VOSTOK AND ZENIT RESPECTIVELY.

1959

JANUARY 5
OFFICIAL APPROVAL GIVEN FOR THE VOSTOK AND ZENIT PROGRAMMES.

1960

MAY 15
THE FIRST VOSTOK SPACECRAFT IS TEST FIRED BUT IS ONLY A PARTIAL SUCCESS.

AUGUST 19
AFTER TWO FAILURES, THE FIRST SUCCESSFUL VOSTOK TEST FLIGHT IS MADE.

1961

MARCH 25
THE LAST UNMANNED TEST FLIGHT OF A VOSTOK CLEARS THE WAY.

APRIL 12
YURI GAGARIN MAKES THE WORLD'S FIRST HUMAN SPACE FLIGHT, LASTING 1HR 48MIN.

MERCURY FLIES!

Training for space

Alongside the beach at Cape Canaveral, Pad 14 supported the Atlas flights carrying Mercury spacecraft on tests. Converted ballistic missile, they were fuelled by kerosene, a form of paraffin, and liquid oxygen. RIGHT • Chimpanzee Ham is made readied for his Mercury flight on a Redstone booster, pioneering the way for human pilots.

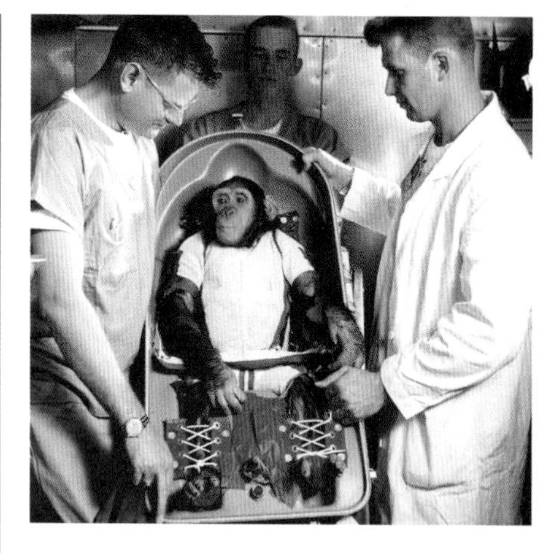

ABOVE RIGHT • *A victim of discrimination, 'Jerrie' Cobb was found suitable for space flight but was prevented from consideration by a law that prohibited women becoming Air Force test pilots, a qualifying criteria for the Mercury programme.* INSERT BOTTOM LEFT • *Malcolm Scott Carpenter undergoes high-g accelerations in a centrifuge at the Navy Air Development Center, Johnsville, Penn. Still the largest ever built, it could exert a rate change of 10g per second and spin up to a simulated 40g load.*

Milestones

1960

FEBRUARY 15:
ASTRONAUTS BEGIN THREE WEEKS OF TESTING ON THE MASTIF TRAINING DEVICE.

1961

JANUARY 31:
CHIMPANZEE 'HAM' WAS BOOSTED TO A HEIGHT OF 157 MILES BY A REDSTONE ROCKET, EXPERIENCING ALMOST 15G DURING DESCENT.

MARCH 24:
A REDSTONE ROCKET IS LAUNCHED TO MAKE THE FINAL TEST FLIGHT BEFORE A PILOTED SUB-ORBITAL HOP.

When NASA selected its astronauts there was no way of knowing what effect space would have on the human body, so they were put through every conceivable test imaginable, for fitness, resilience and the ability to handle stressful situations – which usually meant unpredictable, life-threatening events!

The monstrous MASTIF

It is said that those who control the operations of simulators are among a particularly sadistic group of people. Then again, that's their job. Sadistic too perhaps are those who design fiendish devices to test every uncomfortable sensation from disorientation to induced nausea. So it was with MASTIF – the Multiple Axis Space Test Inertia Facility – comprising three tubular cages set to revolve separately inside each other, like coupled gyroscopes. Motion could be controlled with nitrogen gas jets like thrusters operated by a stick for the occupant's right hand. With the astronaut fixed rigidly in the centre of the cage, the supervisor would start it up with multiple motion in roll, pitch and yaw of up to 30 rpm and monitor the speed with which the 'pilot' grasped the correct thruster motions necessary to bring it to a stop. Simulating a tumbling capsule in orbit, the occupant would neutralise his own motion inside the cage and achieve a stable attitude. It was used by all the Mercury 7 in early 1960, providing data about their performance for up to five hours at a timeas they learned how to use hand control inputs to change the attitude of their spacecraft.

The Mercury '13'

A consultant to NASA on astronaut selection, William Lovelace recruited a woman, Gwendolyn 'Jerrie' Cobb, to undergo all the rigorous tests laid down for the selected seven male Mercury pilots. At this time women were barred from Air Force schools and so could not qualify as military pilots, neither could they apply to be test pilots for the Air Force and were thus unable to qualify as astronaut candidates. Lovelace found the privately funded results encouraging and recruited 18 more women, finding 12 like Cobb who were exceptionally well suited to space flight. Some had more flying experience than the Mercury 7 astronauts, although not on fast jets. Resistance to the idea prevented them from being considered and not until 1978 did NASA select women astronauts.

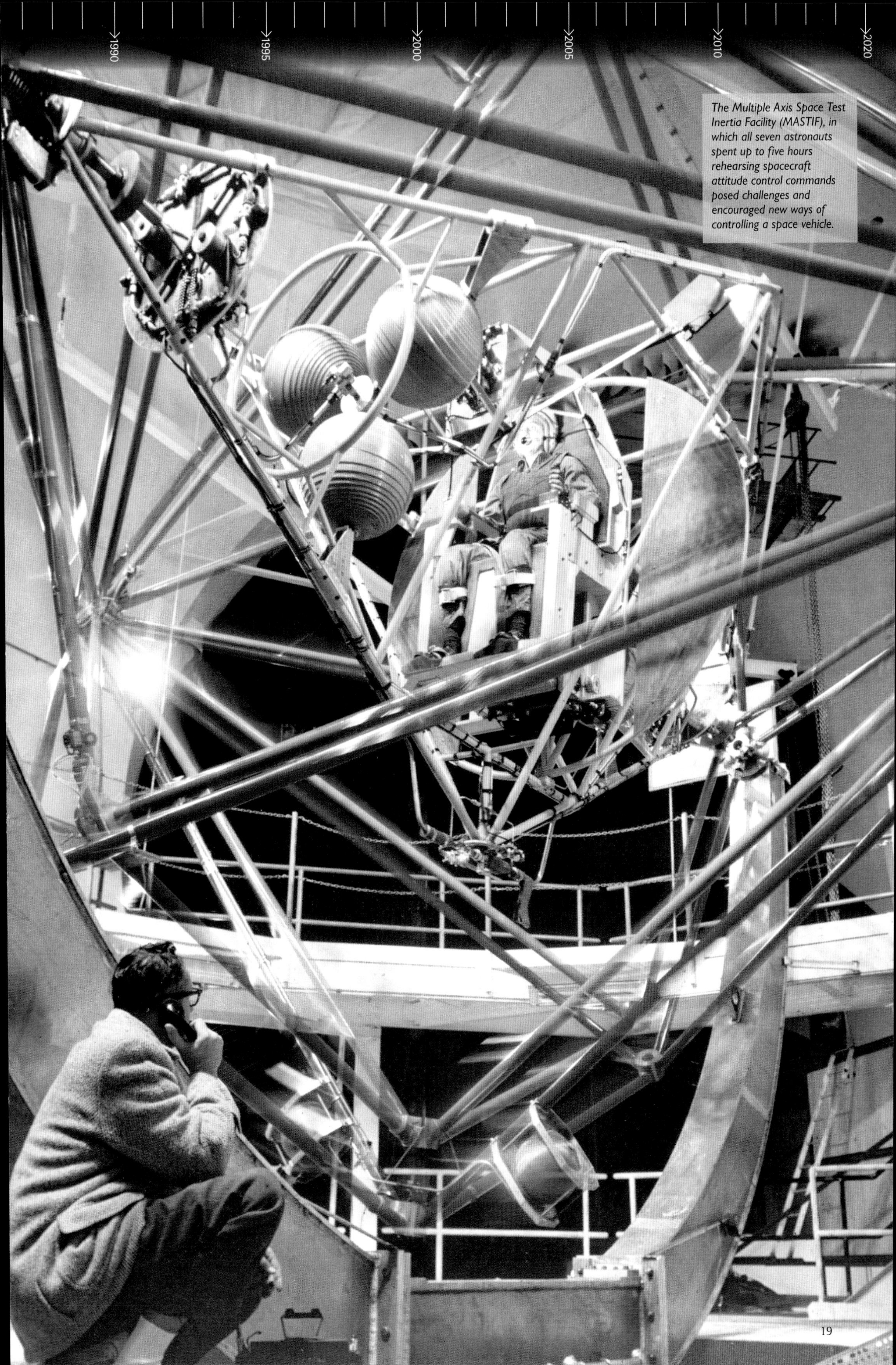

The Multiple Axis Space Test Inertia Facility (MASTIF), in which all seven astronauts spent up to five hours rehearsing spacecraft attitude control commands posed challenges and encouraged new ways of controlling a space vehicle.

THE FIRST AMERICAN IN SPACE

It had taken more than three years but all was finally ready for the first flight of an American astronaut when news broke of Yuri Gagarin's epoch-making mission. Bitter disappointment that NASA had not managed to pull off this most coveted space 'first' ran throughout the space agency but the Mercury programme was about to deliver – albeit with a simple ballistic flight straight and down again to test man and machine.

Launched in the early afternoon of May 5, 1961, Alan Shepard rode the Redstone rocket for 2 minutes 22 seconds before it burned out and the capsule, named Freedom 7, separated. Peaking at an altitude of 116.5 miles and a speed of 5,134 mph, it fell back and began its descent through the earth's atmosphere, imposing 11g on its occupant. It was bobbing in the Atlantic Ocean just 15 minutes 22 seconds after liftoff.

With a thrust of 78,000lb, the Mercury Redstone MR-3 mission gets under way from Pad 5 at Cape Canaveral, May 15, 1961. Grissom and Glenn had also prepared for MR-3 but Shepard was selected by manned flight boss Robert Gilruth two days before the flight. INSERT BOTTOM LEFT • *The suit worn by Mercury astronauts was a custom-fitted adaptation of the US Navy Goodrich Mk IV high altitude jet aircraft pressure suit. It was worn only as a backup to the pressurized capsule.* BELOW • *President Kennedy presented the NASA Distinguished Service Medal to Shepard in the White House Rose Garden on May 8. What the world would not know until May 25 was that Kennedy had already decided to announce a moon race with the Soviets.*

Milestones

1961

APRIL 25: THE MERCURY ATLAS MA-3 FLIGHT FAILS TO REACH ORBIT WITH A TEST CAPSULE.

APRIL 28: A LITTLE JOE BOOSTER FINALLY PROVES THE ABILITY OF THE LAUNCH ESCAPE SYSTEM TO WORK PROPERLY.

MAY 3: ALAN SHEPARD SELECTED AS THE PILOT FOR THE FIRST MERCURY MISSION.

MAY 5: MR-3 CARRIES SHEPARD ON A 15 MINUTE FLIGHT 116 MILES ABOVE THE EARTH FROM WHICH HE IS SAFELY RECOVERED.

MAY 8, SHEPARD IS AWARDED THE NASA DISTINGUISHED SERVICE MEDAL BY PRESIDENT KENNEDY.

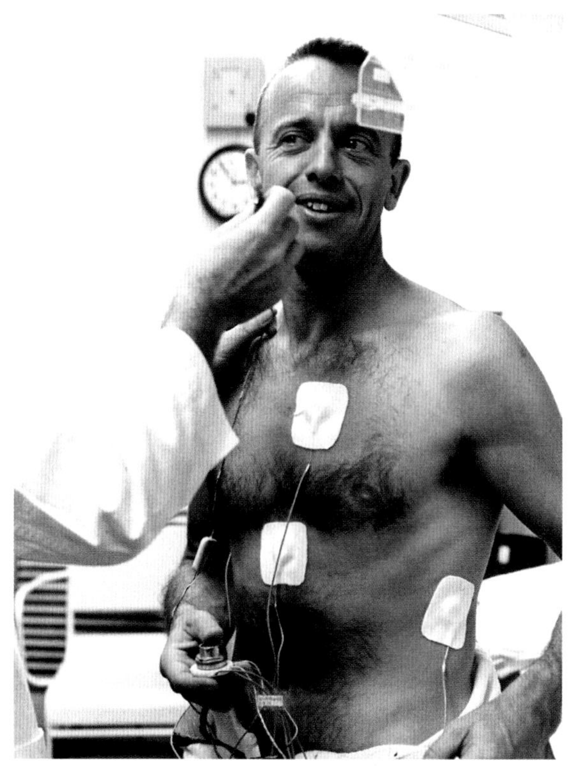

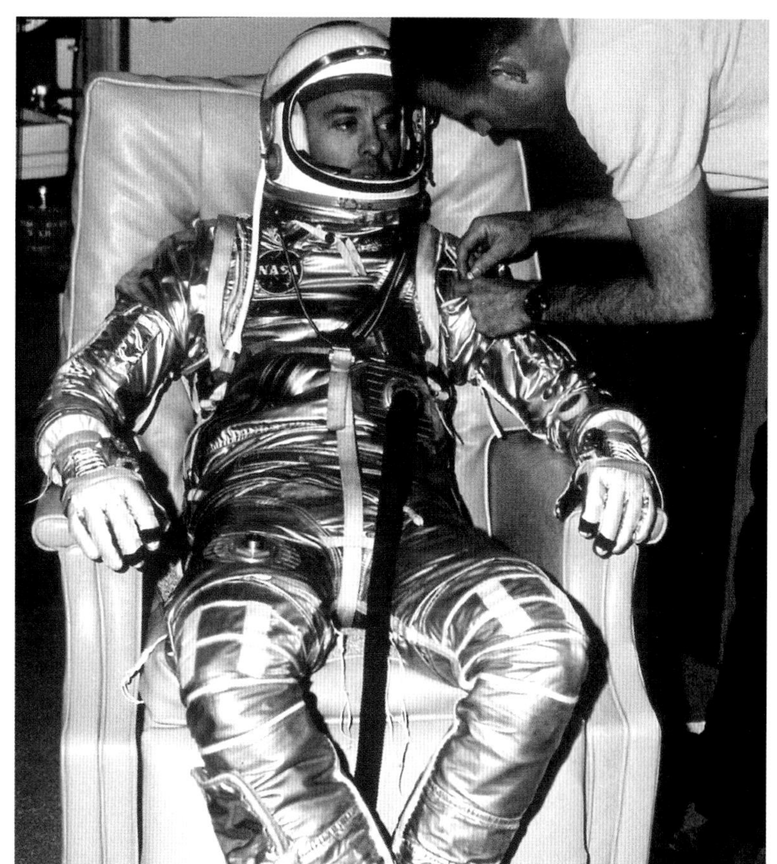

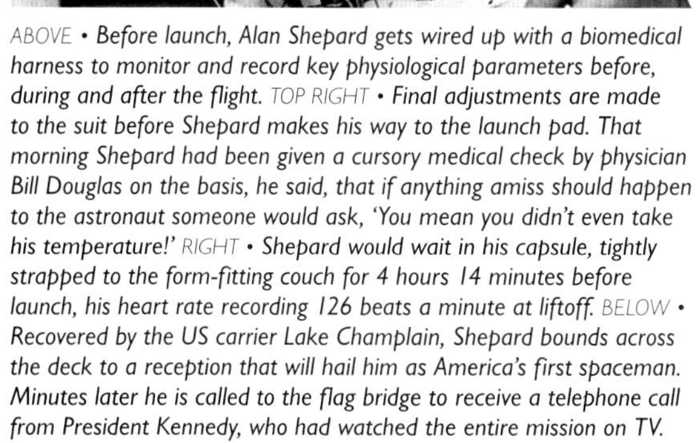

ABOVE • *Before launch, Alan Shepard gets wired up with a biomedical harness to monitor and record key physiological parameters before, during and after the flight.* TOP RIGHT • *Final adjustments are made to the suit before Shepard makes his way to the launch pad. That morning Shepard had been given a cursory medical check by physician Bill Douglas on the basis, he said, that if anything amiss should happen to the astronaut someone would ask, 'You mean you didn't even take his temperature!'* RIGHT • *Shepard would wait in his capsule, tightly strapped to the form-fitting couch for 4 hours 14 minutes before launch, his heart rate recording 126 beats a minute at liftoff.* BELOW • *Recovered by the US carrier Lake Champlain, Shepard bounds across the deck to a reception that will hail him as America's first spaceman. Minutes later he is called to the flag bridge to receive a telephone call from President Kennedy, who had watched the entire mission on TV.*

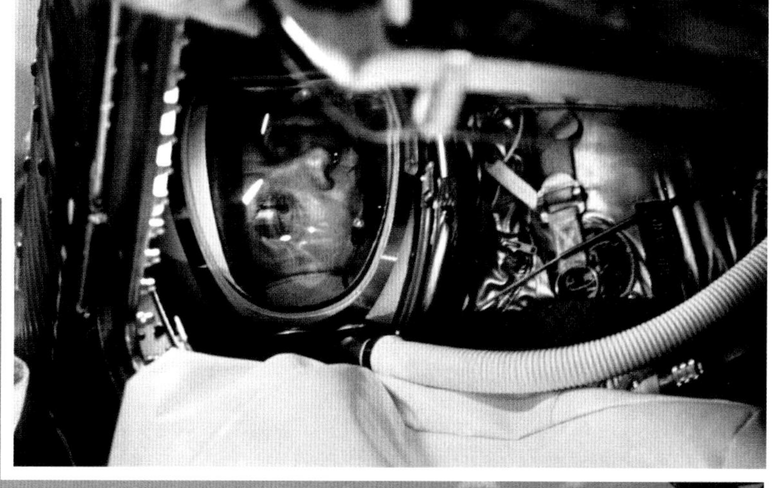

In orbit, a suspect warning light indicated to flight controllers on the ground that the heat shield had come loose. Glenn could control the rotation of his spacecraft in attitude but Mercury had no means of changing its orbit, other than firing the retro-rockets for slowing down to re-enter the atmosphere.

INSET BELOW • An onboard camera records Glenn's attentive glance at instrumentation during his three orbits of the earth.

A final rehearsal

Milestones

1961

JULY 21: CARRYING ASTRONAUT GUS GRISSOM, MERCURY SPACECRAFT LIBERTY BELL 7 IS LAUNCHED ON MR-4, THE SECOND BALLISTIC FLIGHT.

NOVEMBER 29: AN ATLAS BOOSTER LAUNCHED CHIMPANZEE ENOS ON THE FINAL QUALIFYING TEST BEFORE AN ASTRONAUT ATTEMPTS ORBITAL FLIGHT.

1962

FEBRUARY 20: JOHN GLENN BECOMES THE FIRST AMERICAN TO ORBIT THE EARTH, MAKING THREE CIRCUITS OF THE PLANET IN 4 HR 55 MIN.

FEBRUARY 26: IN POURING RAIN, JOHN GLENN IS PARADED ON A MOTORCADE THROUGH WASHINGTON DC AND DELIVERS AN ADDRESS TO A JOINT SESSION OF CONGRESS.

NASA wanted at least one more suborbital hop before attempting the first orbital flight and that took place on July 21, 1961 when Virgil 'Gus' Grissom repeated the flight of MR-3. It was similar to the flight Shepard had taken, except that the hatch jettison mechanism was activated prematurely and the capsule, named Liberty Bell 7, filled with water.

Struggling to free himself Grissom almost drowned, his suit filling with water, while the retrieval helicopter was intent on saving the capsule from sinking. A false signal indicated the helicopter's engine was overheating and it let go of the Mercury, which sank to the bottom of the Atlantic. Liberty Bell 7 was recovered from the ocean floor on July 20, 1999 by an expedition financed by the Discovery Channel and it is now in the Kansas Cosmosphere and Space Center.

Mission accomplished

The flight of Friendship 7 brought the culmination of NASA's ambition to get an American in orbit. That attempt began mid-afternoon on February 20, 1962, and carried John Glenn to a speed of 17,500 mph and an orbit of 100 miles by 162 miles just 5 minutes after liftoff. At 360,000lb, the Atlas delivered more than four times the thrust of the Redstone rocket used for the two previous suborbital shots.

Almost immediately Glenn settled in to a routine of observations, measurements and tests using the hand controller to change the attitude of his capsule. All went well until a faulty warning signal indicated to flight controllers on the ground that the circular heat shield upon which his very survival depended during re-entry may have come loose. It was designed to separate after re-entry and hang beneath the spacecraft as it neared splashdown to cushion the impact of splashdown. Had it come loose in flight after the retro-pack had fired, the shield could have torn free during re-entry.

A degree of doubt

Nobody on the ground knew if the signal was faulty and as a precaution they advised Glenn to leave the retro-pack on instead of jettisoning it after it fired to slow him down. If the shield was loose its straps would hold the shield tight until atmospheric pressure would secure it against the base of the spacecraft. Glenn knew nothing of this until given instructions not to release the pack.

Right on time, toward the end of the third orbit as Friendship 7 was crossing the Pacific, the retro-rockets fired and Glenn began his descent. As heat built up flaming chunks of retro-pack began to break off and hurtle past his window. All the way down he watched as debris shot into view, unsure of whether it was the pack or the shield itself breaking up.

Finally, the capsule slowed, the sky became blue and the parachutes came out delivering him to a watery splashdown in the Atlantic Ocean. It had been a close call. The switch had given a faulty reading but all could have gone horribly wrong. It had not gone wrong, and America had its hero.

Mercury Redstone 4 launches Gus Grissom on a suborbital hop lasting 15 minutes 37 seconds, boosting him to a height of 118 miles.

BELOW • Named Friendship 7, Glenn's spacecraft has a trapezoidal window in the side, unlike Shepard's Mercury spacecraft which had a circular window.

John Glenn rides an Atlas rocket into space on February 20, 1962, the first American to orbit the earth.

BELOW • President Kennedy addresses NASA employees at the Manned Spacecraft Center, then at the Langley Research Center, Virginia. It moved to Houston in 1964.

RIGHT • *Gordon Cooper walks from his spacecraft after being lifted in his capsule and placed on the deck of the Kearsarge. Physicians spent eight minutes examining him for after effects of his 34 hr mission before he was allowed out.*

INSET BELOW • *Carpenter examines the base of his Mercury spacecraft to which the slightly convex, disc-shaped heat shield will be attached.*

Milestones

1962

MAY 24:
MALCOLM SCOTT CARPENTER MAKES THE SECOND ORBITAL FLIGHT (MA-7) IN MERCURY SPACECRAFT AURORA 7, LASTING 4 HR 56 MIN.

OCTOBER 3:
WALTER M 'WALLY' SCHIRRA MAKES A SIX-ORBIT MISSION (MA-8) IN SIGMA 7, LASTING 9 HR 13 MIN.

1963

MAY 15:
GORDON COOPER FLIES THE LAST MERCURY MISSION (MA-9) IN FAITH 7, MAKING 22 ORBITS IN A FLIGHT LASTING 34 HR 20 MIN.

JUNE 6:
SPACE OFFICIALS BRIEF NASA BOSS JAMES WEBB ON FURTHER MERCURY FLIGHTS BUT WITH GEMINI AND APOLLO IN FULL SWING, THE PROGRAMME IS CANCELLED.

A less than glittering performance...

A repeat of Glenn's flight carried Malcolm Scott Carpenter into space in late May, 1962 in a spacecraft named Aurora 7. Deke Slayton had been named as the pilot but doctors detected a mild heart murmur and he was replaced by Carpenter. During the flight Carpenter used too much thruster fuel trying to rotate his capsule around in space and he was forced to turn off the attitude control system for 1 hour 17 minutes to be sure he had enough fuel to control the attitude of his capsule on re-entry.

During that nail-biting event he ran out of thruster gas and the spacecraft began to oscillate before air pressure held it in the correct attitude. Moreover, Carpenter had been late in firing the retro-rockets and splashed down 250 miles off target. He waited three hours, bobbing around on the Atlantic Ocean in his inflatable life-raft until recovery forces arrived. Little more than a year later Carpenter took leave of absence to work on underwater habitats for the Navy and never returned.

Live on TV

Walter M 'Wally' Schirra got to spend six orbits around the earth after launch in October 1962, putting up a stunning performance, saving large quantities of attitude control gas due to an economical use of the spacecraft's thrusters. The entire flight went text-book perfect and Schirra splashed down within 5 miles of the waiting carrier, the USS Kearsarge, in full view of millions of watching Europeans, seeing a mission for the first time live on TV via the Telstar satellite. Schirra came down 275 miles northeast of Midway Island, the first capsule recovered from the Pacific Ocean.

Asleep in space

A NASA plan to keep an astronaut in orbit for more than a day came to fruition with the last Mercury flight launched in May, 1963. Piloted by Gordon Cooper, the flight had few troubles but Cooper did get the chance to take the first full sleep in space. Despite problems with attitude alignment, it came down within four miles of the recovery carrier, USS Kearsarge, again off Midway, more than 34 hours after launch. Some engineers wanted to fly another Mercury mission on an even longer flight but with the two-man Gemini missions looming up there was neither time nor resources to do that. Coming up was the grand plan to head for the moon and before that, the two-man Gemini missions promised space spectaculars unimagined when the Mercury programme began in October 1958.

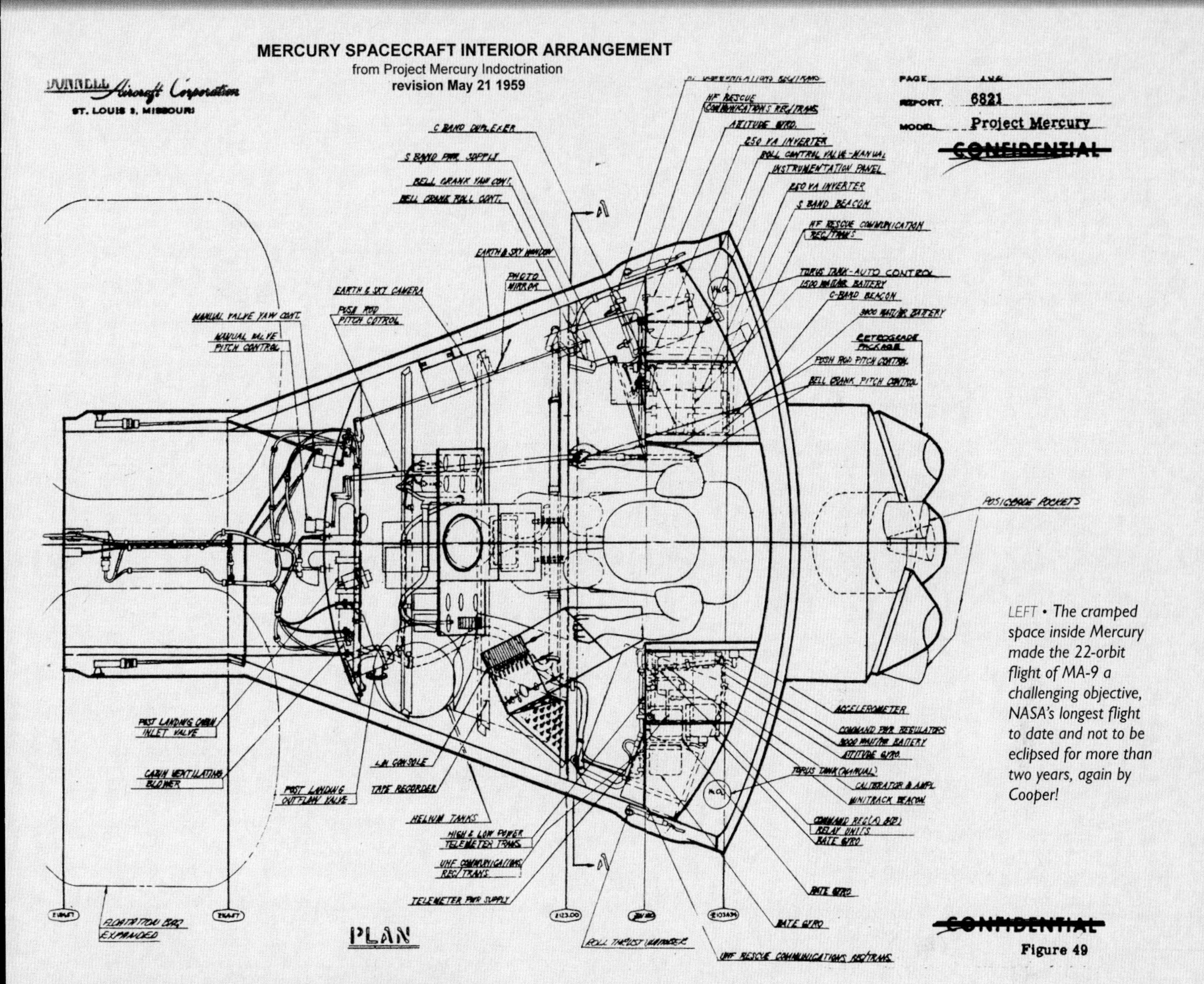

LEFT • The cramped space inside Mercury made the 22-orbit flight of MA-9 a challenging objective, NASA's longest flight to date and not to be eclipsed for more than two years, again by Cooper!

FAR LEFT • To prepare for the unforeseen eventuality of having to make an emergency landing on earth, astronauts learned how to survive in remote locations and received training in desert and jungle survival.

LEFT • The launch of MA-7, NASA's second Mercury orbital flight from Pad 14 at Cape Canaveral, May 24, 1962.

FAR LEFT • Divers and sailors from the USS Kearsarge attach flotation collars around the Mercury 9 spacecraft, named Faith 7.

TO THE MOON!

Challenging the Russians

With a thrust of 1.3 million lb, four times the power of the Mercury-Atlas rocket, the first Saturn I launch vehicle is successfully launched from Cape Canaveral on October 27, 1961.

FAR RIGHT • *The first stage of a giant Saturn V rocket, more than five times the power of the Saturn I and 20 times more powerful than the Atlas used in Mercury flights.*

BELOW • *V-2 design chief Wernher von Braun had been working for the US Army since 1946. He had developed the Jupiter C rocket that placed the first US satellite in orbit in 1958 and the Redstone that sent the first American into space in 1961. Now he was in charge of the Saturn rockets that would pave a path to the moon.*

INSET RIGHT • *The Lunar Landing Research Vehicle, or LLRV, was developed to give astronauts experience with flying a vehicle using thruster jets, while the main weight of the contraption was supported by a downward firing jet engine.*

RIGHT • *Top NASA managers give President Kennedy an update on Saturn in the blockhouse to launch pad 34 at Cape Canaveral, September 1962. From left, NASA boss James Webb, Vice President Lyndon Johnson, Launch Director Kurt Debus and President John F Kennedy.*

Outgoing President Eisenhower had been accused of letting the Russians beat America to putting the first satellite in space but in early 1961 the newly elected President Kennedy had his own agenda, challenged when Yuri Gagarin became the first man in space

The unwanted spacecraft

President Eisenhower set up NASA to accelerate America's space plans after the embarrassment of Sputnik 1, but in 1958 it had been Senate majority leader Lyndon Johnson that organized the emergence of the new agency for space. In the last year of office, however, Eisenhower had stopped NASA developing Mercury's successor, which in July 1960 had been named Apollo after the Greek god representing light and truth.

When Kennedy became President in January 1961, he too was lukewarm to the idea of a successor to Mercury but when Gagarin was launched, embarrassed by worldwide acclaim for the Soviet triumph, he asked his Vice President, Lyndon Johnson, to find a way of beating Russia in space. Johnson was the architect behind the plan to turn a simple successor to Mercury into a moon-landing goal.

...Before the end of the decade...

Suddenly, Apollo seemed the only way to throw a challenge so bold that the Russians just might not be able to accomplish it. On May 25 President Kennedy appeared before a Joint Session of Congress to utter the historic challenge: I believe that this nation should commit itself to achieving the goal, before this decade is out, of landing a

man on the moon and returning him safely to the earth'.

Ironically, Kennedy never said he wanted to do it before the Russians, but everybody who heard those words took it as an undeclared challenge, the declaration of a Cold War race in space whose rules would be set not by the Soviet Union but by the United States.

Building it big

Apollo would need rockets and spacecraft bigger than anything built before but much of it was already on the drawing boards. Former German V-2 rocket scientist Wernher von Braun was developing the Saturn rockets at NASA's Marshall Space Flight Center and the specification for the Apollo spacecraft had already been written. Added in 1962 would be a Lunar Module to go down from moon orbit, where the Apollo spacecraft would remain, to the surface. Not for 18 months would the precise method be finalized but the race was on and the pace quickened.

RIGHT • *As developed, the Lunar Module had upper (ascent) and lower (descent stages), the lower section remaining on the moon as a launch pad for the upper section carrying the crew back to the Apollo mothership in lunar orbit.*

Accompanied by NASA's policy chief Robert Seamans (left), Wernher von Braun explains the workings of a Saturn rocket to President Kennedy.

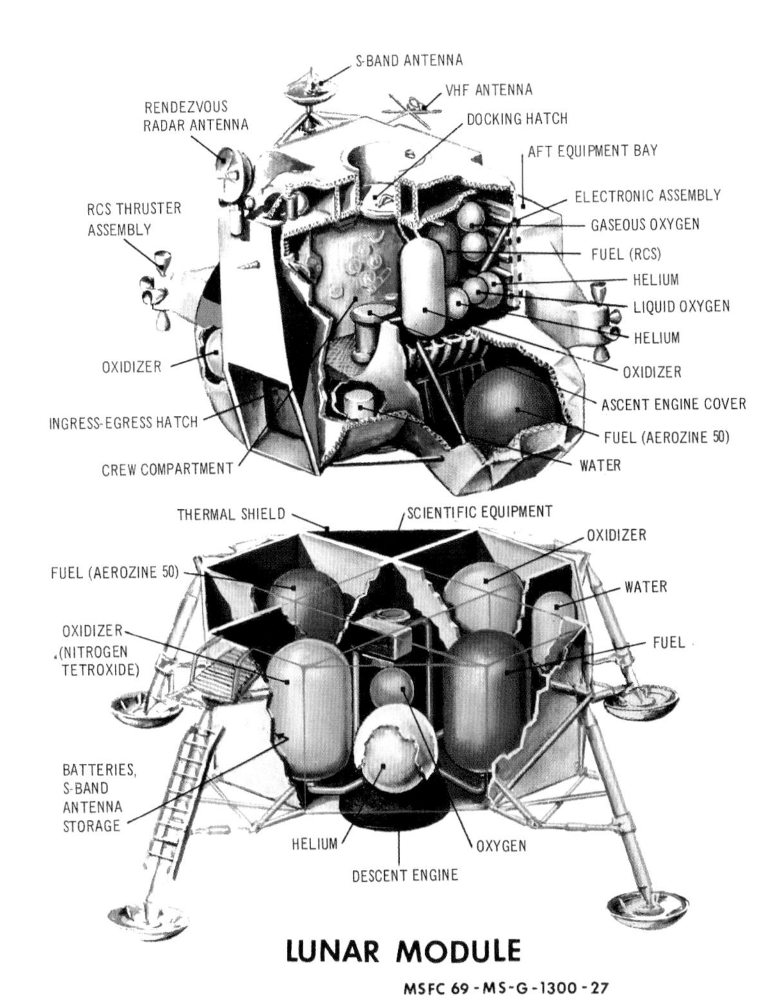

BELOW • *The early version of Apollo envisaged the entire spacecraft descending backwards to the surface of the moon, using landing legs to keep it upright.* BELOW RIGHT • *An early design concept for the Lunar Module, developed as a separate spacecraft to put men on the moon.*

SOVIETS SURGE INTO SPACE

Above: Muscovites need little excuse to party but their excitement over the three-man flight of Voskhod I was genuine. For Russians the space exploits of their cosmonauts became regular cause for celebration.

Milestones

1961

AUGUST 6: VOSTOK 2 LAUNCHED WITH GHERMAN TITOV ON A FLIGHT LASTING 25HR 18MIN.

1962

AUGUST 11: ANDRIAN NIKOLAYEV LAUNCHED IN VOSTOK 3 FOR A 3 DAYS, 22HR 22MIN MISSION.

AUGUST 12: VOSTOK 4 CARRIES PAVEL POPOVICH INTO ORBIT ON THE FIRST DUAL FLIGHT MISSION, LASTING 2 DAYS, 22HR 56MIN.

1963

JUNE 14: VALERY BYKOVSKY LAUNCHED ABOARD VOSTOK 5, AT 4 DAYS, 23HR 7MIN, THE WORLD'S LONGEST SOLO FLIGHT.

JUNE 16: VALENTINA TERESHKOVA, THE FIRST WOMAN COSMONAUT ABOARD VOSTOK 6, LASTING 2 DAYS 22HR, 50MIN.

1964

OCTOBER 12: VOSKHOD I CARRIES KOMAROV, FEOKISTOV AND YEGOROV ON THE FIRST MULTI-MAN FLIGHT IN A HEAVILY ADAPTED VOSTOK CAPSULE, A FLIGHT LASTING 24HR 17MIN.

Not content with putting the first man in orbit, the Russians now began to widen the gap with a new round of 'firsts'.

Sergei Korolev knew that his Vostok spacecraft was capable of much more than a single orbit of the earth and proved that in August, 1961, with the launch of Gherman Titov on the first manned mission to last a full day in space. Still the Americans had not sent a man into orbit and still the Russians were claiming new leads in space. Yet Vostok 2 was but precursor to a string of successful flights with this spacecraft over the next two years.

A year after Titov's flight, two Vostoks were launched a day apart in the first twin mission to space. Vostok 3 carried Andrian Nikoleyev into orbit on August 11, 1962, followed by Pavel Popovich in Vostok 4 on August 12 with just three miles separating them. Together they were monitored by Moscow in a dramatic increase in capabilities already impressive in their flexibility. But it nearly came to high drama when a coded word 'thunderstorms', implying something seriously wrong with the capsule, was transmitted by Popovich as he excitedly informed the ground that he really could see 'thunderstorms' on the earth below!

New firsts for women

Next to fly was Vostok 5 in June 1963 carrying Valery Bykovsky into orbit on what would remain to this day the longest solo mission of all. Before he returned to earth, Valentina Tereshkova, then only 26 years old, became the first woman cosmonaut when she was launched aboard Vostok 6 to within three miles of Bykovsky. Chatting to each other by radio they beamed to earth the first live TV pictures from space and struck a propaganda coup over NASA, which had fiercely opposed sending women into space.

While Tereshkova spent almost three days in orbit, when Bykovsky returned to earth on June 19, 1963, he had been in space for almost five days, a record for a manned vehicle remaining in space that would not be beaten by the US for two years. But the Americans were catching up, the moon challenge had been set and NASA's two-man Gemini was nearing its first test flights. Krushchev wanted more spectacular flights.

A child of the Space Age

Just four months after becoming the first woman in space, in November 1963 Valentina Tereshkova married cosmonaut Andrian Nikolayev from the Vostok 3 mission of August 1962. Their daughter Elena was born in June 1964, the first child born to a couple who had both been into space. But it failed to outlast the publicity and amid unsubstantiated talk that it had all been an experiment to test fertility in each gender their relationship fell apart and they were divorced in 1980.

New risks

Further Vostok missions were cancelled and Korolev modified the basic spacecraft into one for multi-man missions known as Voskhod. Now Russia was scrambling to keep ahead of the Americans and on October 12, 1964, the first three-man flight took place when cosmonauts Komarov, Feoktistov and Yegorov were launched by a more powerful version of the rocket previously used for Vostok flights. Their mission lasted little more than a day and when they returned to earth the Soviet government had a new leader – Nikita Krushchev had been replaced in a coup orchestrated by Leonid Brezhnev.

Voskhod had been a dangerous gamble. By removing the ejection seat and having the cosmonauts wear lightweight clothing three men had been squeezed into a capsule made for one. As a nod to safety, a backup retro-rocket was mounted on the capsule and to prevent broken bones on landing, another retro-rocket was attached for braking the descent seconds before hitting the ground. Now there was to be an even more daring adaptation of the Vostok spacrecraft.

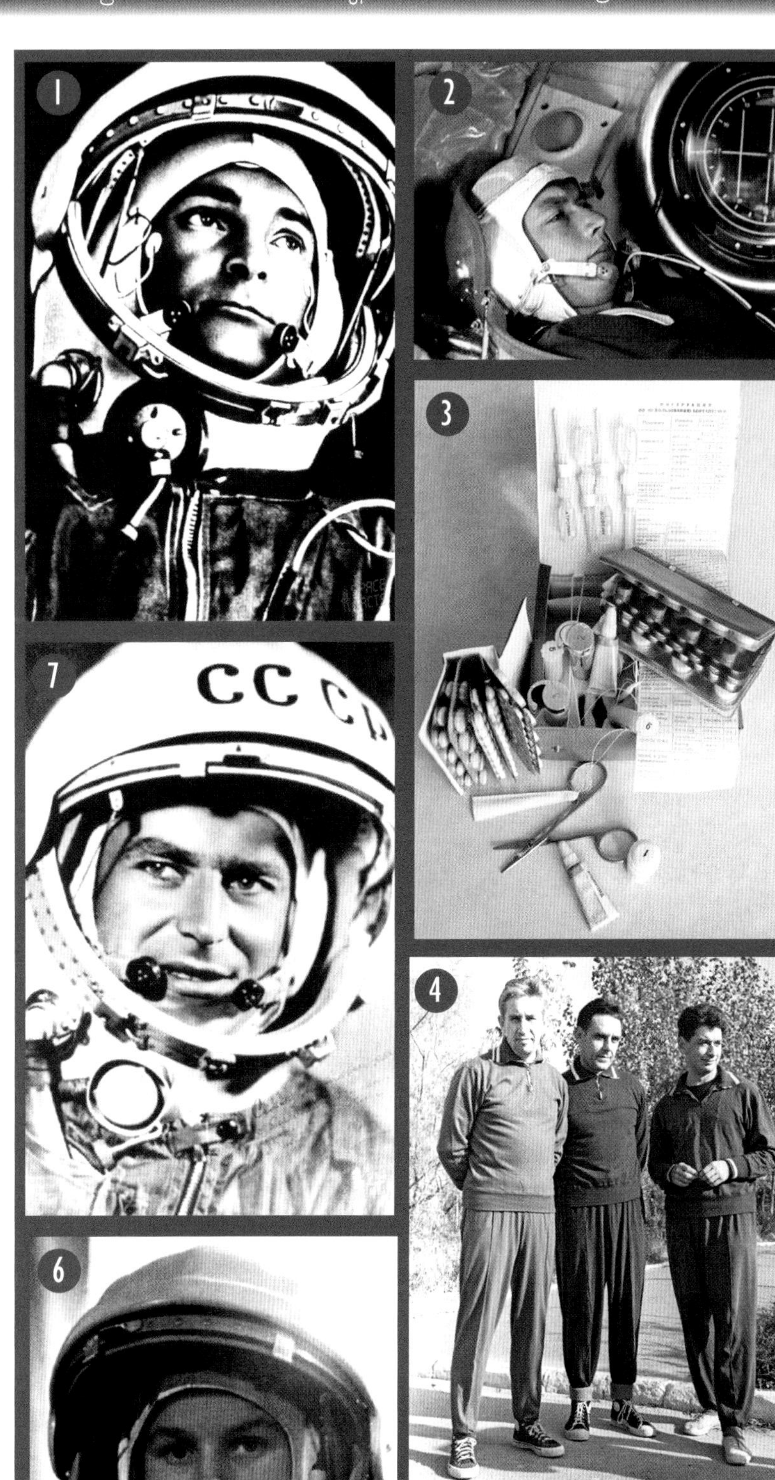

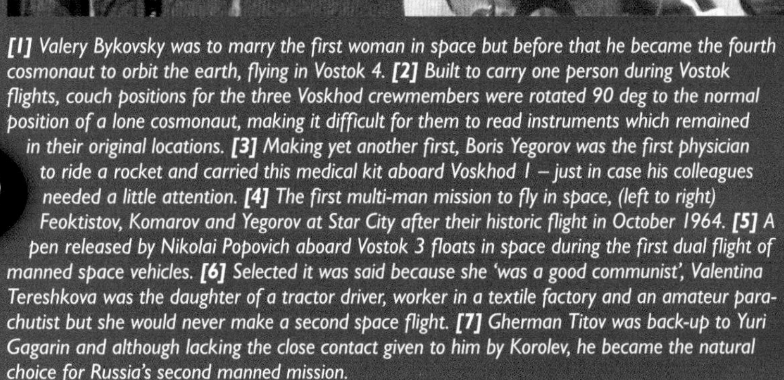

[1] Valery Bykovsky was to marry the first woman in space but before that he became the fourth cosmonaut to orbit the earth, flying in Vostok 4. [2] Built to carry one person during Vostok flights, couch positions for the three Voskhod crewmembers were rotated 90 deg to the normal position of a lone cosmonaut, making it difficult for them to read instruments which remained in their original locations. [3] Making yet another first, Boris Yegorov was the first physician to ride a rocket and carried this medical kit aboard Voskhod 1 – just in case his colleagues needed a little attention. [4] The first multi-man mission to fly in space, (left to right) Feoktistov, Komarov and Yegorov at Star City after their historic flight in October 1964. [5] A pen released by Nikolai Popovich aboard Vostok 3 floats in space during the first dual flight of manned space vehicles. [6] Selected it was said because she 'was a good communist', Valentina Tereshkova was the daughter of a tractor driver, worker in a textile factory and an amateur para-chutist but she would never make a second space flight. [7] Gherman Titov was back-up to Yuri Gagarin and although lacking the close contact given to him by Korolev, he became the natural choice for Russia's second manned mission.

FLOATING ON SPACE

In a final bid to beat the Americans, a Russian cosmonaut steps outside for a walk across the world.

RIGHT INSET: Carrying a false depiction of what Voskhod really looked like, this stamp supports the propaganda message that Leonov's spacecraft was a completely new design.

BOTTOM RIGHT: Essentially a Vostok capsule into which the two suited cosmonauts would squeeze, there was no room for movement with only minimal control over the capsule's functions.

BOTTOM LEFT: Leonov wore a modified Sokol-1 suit previously used on Vostok flights to guard against unexpected depressurization of the capsule, now used for the first space walk.

BELOW: Had the Russians made it to the moon, Leonov would have commanded the first landing. As it was he commanded the Russian Soyuz on the first US-Soviet docking flight in 1975.

The Soviet chief spacecraft designer Sergei Korolev knew the Americans were planning to fly a new class of spacecraft called Gemini. It was to rehearse all the techniques needed for moon missions – including a space walk. That was one goal left over from a string of 'firsts' that kept the Russians way ahead of NASA during the first four years of manned flight. The Americans had flown six Mercury missions, the longest lasting little more than one full day in orbit. The Russians had flown seven missions of up to five days in duration and twice launched dual flights – and they had pre-empted the emergence of a multi-man vehicle with the three-person crew of Voskhod 1. Now was the time to take an even bolder step by setting up the first space walk.

Voskhod 2

Modifying Vostok to adapt it for three people was relatively easy but the way the basic spacecraft could be modified to support space walking brought unique problems. Vostok could not be repressurised in space and so leaving the capsule required an airlock to avoid depressurizing the interior. But the exterior confines of the capsule meant that any extension or additional structure would take time to develop and adapt to the launch vehicle.

The only solution was to carry an inflatable airlock carried on the side of the capsule folded tight against the exterior for launch. It was called Volga and it had a diameter of just under 4ft and a length of 8ft. After reaching orbit Volga would be pressurised with air from inside the capsule, inflating it to full size large enough for a man to open the hatch in the side of the capsule, crawl inside, close it again. After depressurizing Volga, a hatch on the opposite end of the airlock would allow the cosmonaut to emerge.

There would need to be two cosmonauts – one to go out and one to stay inside the capsule – and for this most dangerous of all space missions flown so far two cosmonauts were selected with impeccable training records. Pavel Belyayev and Alexei Leonov had been among the original group of cosmonauts and were in line for a space flight when they were chosen to fly Voskhod 2.

Stepping out

The flight of Voskhod 2 began during the morning of March 18, 1965. Within five minutes the capsule was in space and at the end of the first revolution as the spacecraft came back over the Soviet Union the Volga airlock was inflated and pressurised. Opening the hatch, crawling inside the airlock and closing it again behind him, Leonov waited while Belyayev operated a valve to allow air out of the airlock. At 11.33am Moscow time he opened the hatch and two minutes later had worked his way out of the now limp cylinder attached to the side of Voskhod 2. The first space walk had begun.

For 12 minutes he floated around, set up a camera and marveled at the view of earth laid out before him, tethered by a 16ft long line that carried communications with Belyayev and with the ground. When it came time to get back in he found his suit had become disarranged, his hands had slipped out of the gloves' fingers and his feet had ridden up into the ankle region of the pressurized garment. The stiff suit would not allow him to get back inside the flexible cylinder, everything kept slipping out of his grasp. His life-giving air was running out.

After what seemed hours of struggling, water building up in the suit from exertion and perspiration, he finally managed to struggle inside head first. But that was the wrong way round and curling up as tight as the suit would allow, Leonov twisted around so that he could close the outer hatch allowing Belyayev to inflate the airlock. The only way Leonov had been able to struggle back inside the airlock was to release the pressure valve on his suit allowing it to partially deflate. Seriously at risk of oxygen starvation he only just managed to get back inside the capsule.

When it came time to return to earth the capsule became unstable and started tumbling around so the automatic system was turned off and the crew manually fired the retro-rocket which carried them far away from their planned landing point. Coming down in a snow covered forest region, it was a day before rescuers found them and a further two days before trees could be cut down to allow helicopters to pluck them out for celebrations in Moscow.

What Leonov had accomplished was the last big 'first' for the Soviet manned space programme. Not for a further two years would another cosmonaut fly into space. When he did it would bring a day of mourning to the Russian people.

Milestones

1965

MARCH 18
VOSKHOD 2 IS LAUNCHED WITH COSMONAUTS LEONOV AND BELYAYEV ON THE FIRST SPACE WALK LASTING 12MIN ON A FLIGHT 26HR 2MIN IN DURATION.

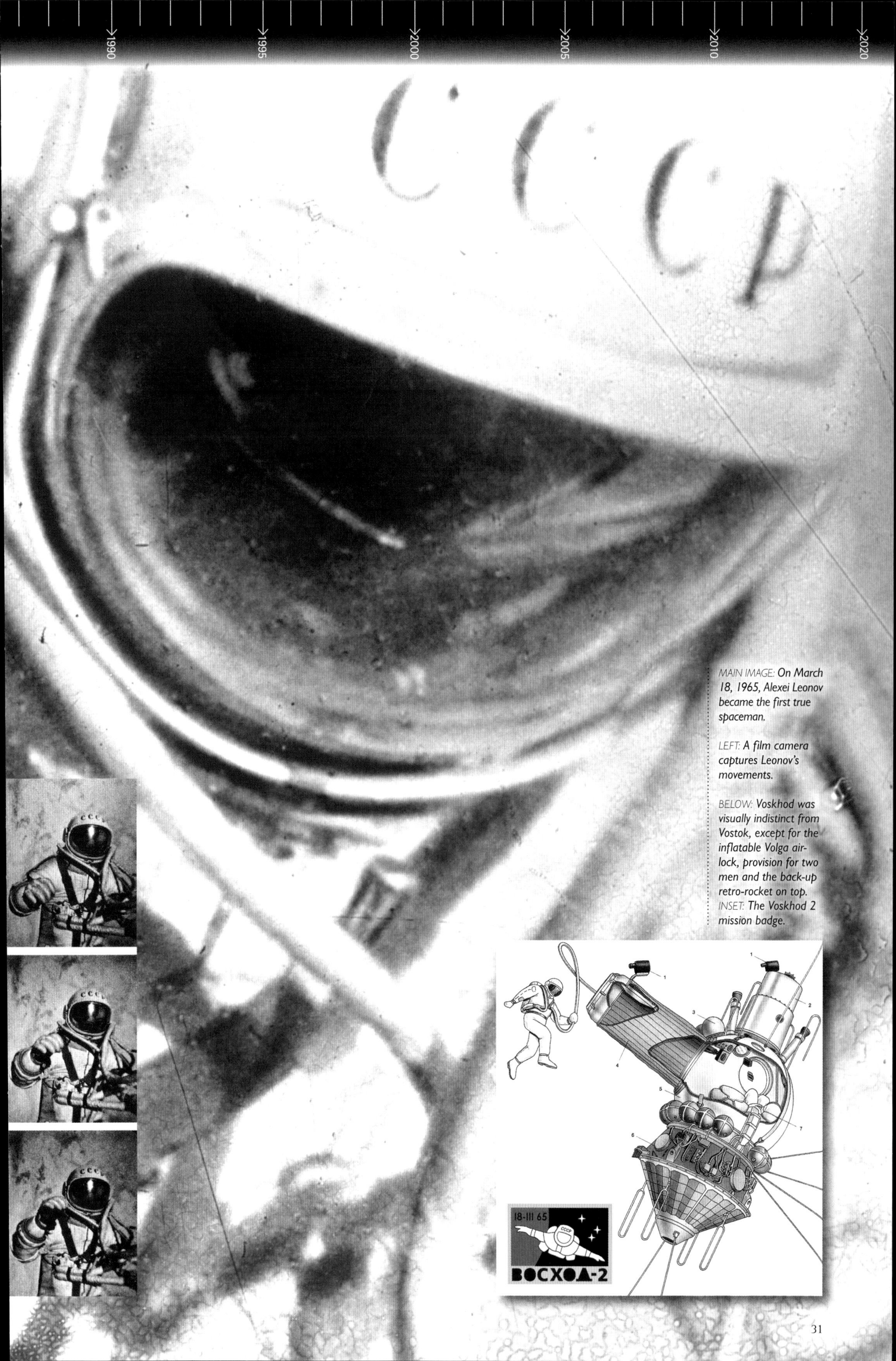

MAIN IMAGE: ***On March 18, 1965,*** Alexei Leonov ***became the first true spaceman.***

LEFT: ***A film camera captures Leonov's movements.***

BELOW: ***Voskhod was visually indistinct from Vostok, except for the inflatable Volga air-lock, provision for two men and the back-up retro-rocket on top.***
INSET: ***The Voskhod 2 mission badge.***

John Young and Gus Grissom in training for the first Gemini mission.

TWINS IN ORBIT

Gemini paves the way

With NASA's moon goal set by President Kennedy in May 1961 there was no time to waste getting ready for the big adventure. With the end of the decade set as the finishing line, and with development of the Apollo spacecraft and the moon landing Lunar Module likely to take several years, the astronauts would need an interim spacecraft in which to practice rendezvous and docking with another object in space, to carry out space walking and to measure the response of the human body to flights lasting up to two weeks.

RIGHT • The Gemini spacecraft was 19ft tall, 10ft in diameter and weighed about 8,000lb. The crew occupiued the dark coloured re-entry module.

In October 1961 McDonnell was selected to upgrade the tiny Mercury capsule into a two-man spacecraft capable of doing all these things and more. Unlike Mercury, where the astronaut was little more than a passenger, able to do few tasks other than control the attitude of his capsule, Gemini would allow its crew to fire rocket thrusters and change orbit, making it possible to catch up with another object in space and dock with it. Just as astronauts in Apollo would have to do, steering their mother-ship back to dock with the Lunar Module after it came up from the surface of the moon.

With a base diameter of 7ft 6in, Gemini provided more room for its crew and carried an adapter in the form of a tapered cylinder, connecting it to the 10ft diameter of the second stage of the Titan 2 rocket that would launch it into space. Developed as a ballistic missile like the Atlas used to lift Mercury into space, Titan was chosen as the only rocket then available with sufficient thrust to place the 8,000lb Gemini spacecraft in orbit. The Saturn rockets, built by Werner von Braun and carrying out test launches at Cape Canaveral, would not be ready for several more years. They were being built for Apollo missions planned to begin in 1967.

Flight tests begin

The Gemini programme was planned to carry two astronauts on each of 10 flights about two or three months apart. But it was late, more development work was needed than NASA had planned for and the first of two unmanned test shots, Gemini-Titan 1, or GT-1, did not fly until April, 1964. Once in space the spacecraft remained attached to the Titan's second stage while instruments measured many aspects of its performance but there was no attempt to recover it. The second test shot was delayed until January 1965 by a string of problems, including a hurricane that sent engineers scurrying to remove it from the pad to a place of safety until the storm blew through! But this time Gemini 2 was recovered from orbit in a full-up test of the on-board systems and the heat shield, the only means of protection back from space.

Milestones

1961

27 OCTOBER: NASA ASKS MCDONNELL TO DESIGN A TWO-MAN VERSION OF MERCURY

1963

21 AUGUST: TITAN 2 ROCKET FOR GEMINI IS CLEARED FOR CARRYING HUMANS TO SPACE

1964

8 APRIL: GEMINI TITAN-1 (GT-1) QUALIFIES THE NEW GEMINI SPACECRAFT

1965

19 JANUARY: THE UNMANNED GT-2 SPACECRAFT IS LAUNCHED AND RECOVERED

23 MARCH: GT-3 IS LAUNCHED WITH GRISSOM AND YOUNG ON A THREE-ORBIT MISSION

3 JUNE: MCDIVITT AND WHITE ON A FOUR DAY MISSION AND NASA'S FIRST SPACEWALK

“ IF ALL WENT WELL, THIS WAS THE YEAR THAT NASA WOULD PULL AHEAD OF THE RUSSIANS - AND BOTH SIDES KNEW IT! ”

Ed Whte performs NASA's first space walk with a gas gun for maneuvering and an emergency oxygen chest pack, June 3, 1965

LEFT • A new group of astronauts was brought on board in 1962 and 1963 for the Gemini and Apollo missions.

Gemini had about as much room as two telephone booths. Note the attitude control stick on the centre console.

Americans back in space

It had been nearly two years since a NASA astronaut had been in orbit but when Grissom and Young soared into space on 23 March 1965, there was a mood of optimism and adventure – a production line spacecraft built to do what no other manned spacecraft had done before. If all went well, this was the year that NASA would pull ahead of the Russians – and both sides knew it!

Grissom and Young spent nearly five hours in space during which they changed the orbit of their spacecraft several times, proving that on future flights they could rendezvous and dock with a previously launched target vehicle. But not just yet, that was planned for Gemini 6.

A Titan 2 rocket lifts the two-man Gemini spacecraft into orbit.

GEMINI EQUIPMENT ARRANGEMENT

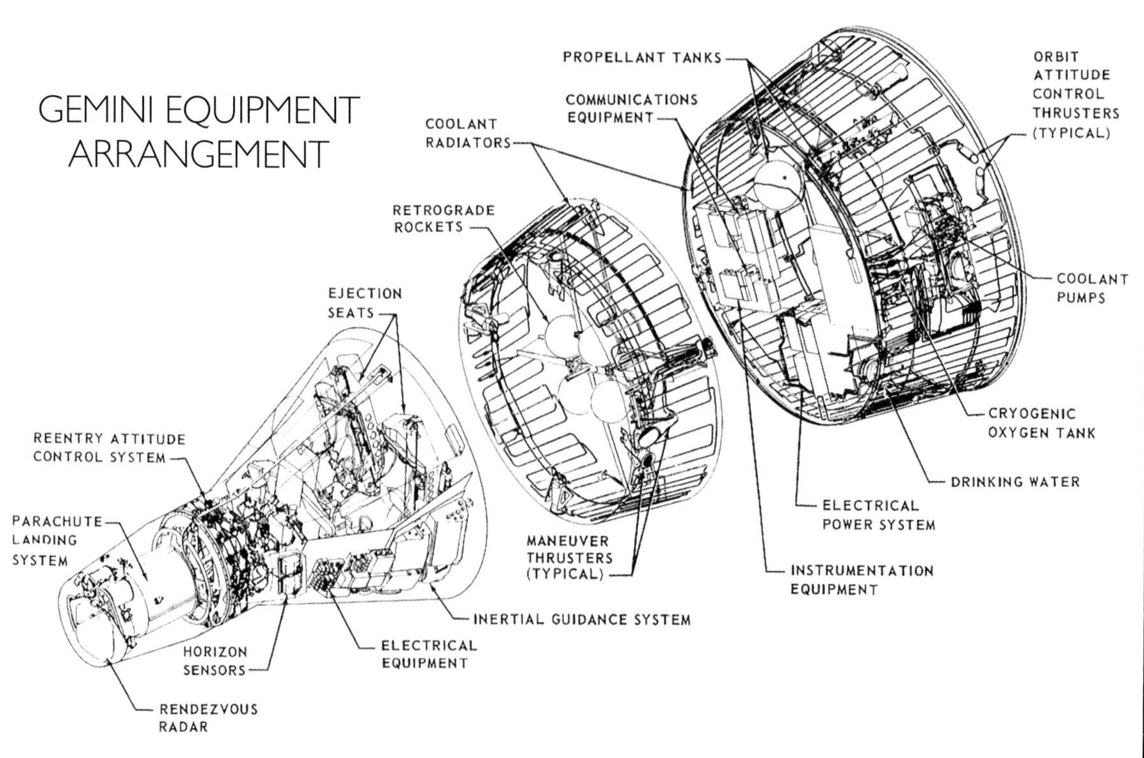

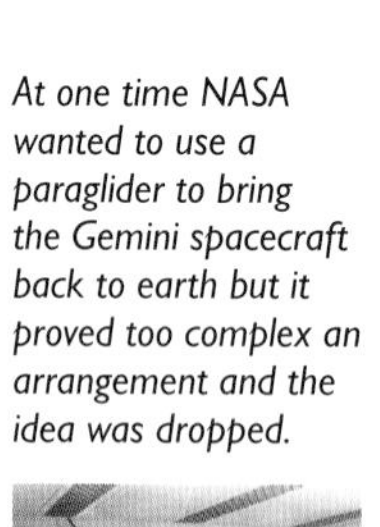

At one time NASA wanted to use a paraglider to bring the Gemini spacecraft back to earth but it proved too complex an arrangement and the idea was dropped.

When they returned to earth after three orbits they cleared the way for the first American to leave his spacecraft and walk in space.

That event took place on June 3, 1965, several hours after the launch of Gemini 4, when Edward H White II became the first American to conduct EVA – Extravehicular Activity, spacewalking. His command pilot was Jim McDivitt and both were rookie astronauts when launched, seasoned spacemen when they splashed down after four days in space, more than the accumulated duration of all previous US manned flights. Their mission was one in a series of incremental steps towards a 14-day flight that would encompass the duration of an Apollo moon mission and prove that humans could survive in weightlessness for that period.

The electric spacecraft

Launched on 21 August 1965, Gemini 5 carrying Mercury astronaut Gordon Cooper and rookie pilot Charles 'Pete' Conrad, thundered away from Launch Complex 19 at Cape Canaveral at the start of an eight-day mission. Theirs was the first spacecraft powered by fuel cells, electricity from hydrogen and oxygen brought together over a nitrogen catalyst with water as a by-product. There were difficulties with getting the fuel cells to work properly but this was a big step into the future.

Apollo would be powered by fuel cells and they simply had to work. The amount of electrical energy required for flights of more than a week or so could not be supplied by any other system. Batteries would be too heavy and in the 1960s solar cells were simply not efficient enough to provide the necessary power.

The two astronauts of Gemini 5 limped along with ailing fuel cells but they stuck to their mission, eked out the power by switching off unnecessary equipment whenever possible and made it back to earth after their scheduled eight days in space. In just three flights NASA had leaped ahead in the technology race with Russia. Now it was time to give Gemini its head!

Astronauts Cooper and Conrad arrive at the launch pad for their eight day flight aboard Gemini 5, the first spacecraft to carry fuel cells.

Stafford (left) and Schirra carry out a suit check before their mission to rendezvous with Gemini 7.

Nearly but not quite!

The next challenge for Gemini was to apply the orbit changing maneuvers tested on earlier missions and rendezvous and dock with an Agena rocket stage, launched from Cape Canaveral one orbit before the astronauts. But if previous flights had seemed smooth, the next phase in this two-man programme was anything but that, as Wally Schirra and Tom Stafford found out when they slipped onto their seats aboard Gemini 6 on October 25, 1965. Waiting aboard Gemini 6 for their turn to launch, the Agena blasted off into space from an adjacent pad at Cape Canaveral. Before reaching orbit the rocket motor failed and Agena plummeted into the ocean.

Within days, NASA came up with an audacious idea. Instead of using up their limited supply of Agena target vehicles, proceed with the next mission, Gemini 7 in December, and send Gemini 6, renamed Gemini 6A, after it for a rendezvous but no docking. Because Gemini 7 was due to remain in space for 14 days, there would be time enough to stack the Gemini-Titan rocket on the same pad and launch it before GT-7 came home. In that way, astronauts would get experience with rendezvous

"IT WAS A SPACE BALLET OF THE MOST EXQUISITE SORT AS THE TWO SPACECRAFT PIROUETTED AROUND EACH OTHER..."

RIGHT AND BELOW • *Gemini 7 viewed from Gemini 6A.*
INSET • *The Andes viewed from Gemini 7.*

Milestones

1965

21 AUGUST: COOPER AND CONRAD REMAIN IN SPACE FOR EIGHT DAYS AND FLY ON FUEL CELLS

25 OCTOBER: SCHIRRA AND STAFFORD FAIL TO LAUNCH IN GT-6 WHEN TARGET VEHICLE CRASHES

4 DECEMBER: GEMINI 7 LAUNCHED WITH BORMAN AND LOVELL ON 14 DAY MISSION

12 DECEMBER: GT-6A ABORTS AFTER ENGINE IGNITION ON THE PAD

15 DECEMBER: SCHIRRA AND STAFFORD LAUNCH ABOARD GT-6A AND CHASE DOWN GT-7

17 DECEMBER: GEMINI 6A COMES HOME AFTER A SUCCESSFUL RENDEZVOUS

18 DECEMBER: GEMINI 7 SPLASHDOWN AFTER 14 DAYS IN SPACE.

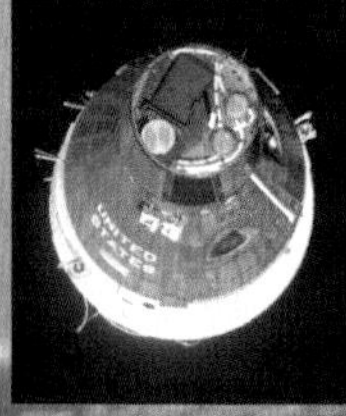

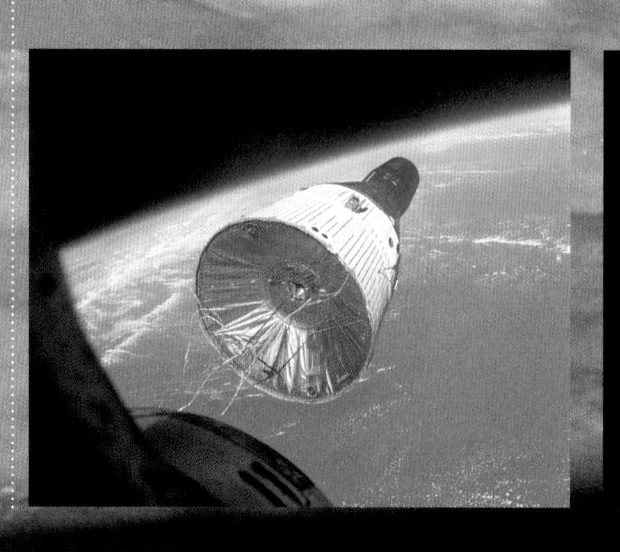

and with the delicate handling of two manned spacecraft in close proximity to each other in space. The docking bit could be left to the next mission, Gemini 8 and its assigned Agena target vehicle.

Gemini 7 was launched on 4 December and the first attempt to send Gemini 6A up came eight days into 7's two-week mission. But just 1.6 seconds after ignition of the Titan booster motors they shut down and the stack settled back on to its cradle! Gemini had two ejection seats in case of a disaster on the pad or on the way up but if Schirra had pulled the abort handle it would have prevented a second launch attempt. The Titan could easily have blown up, but Schirra and Stafford stayed put and lived to launch another day.

Four men in space

With Gemini 7 already 11 days into its 14 day mission, Gemini 6A finally got off the pad at the third attempt on 15 December. In a magnificent sequence of rendezvous maneuvers the astronauts chased down their target in space, closing the 1,238 miles that separated the two spacecraft at orbit insertion to just a few feet less than six hours after launch. It was a space ballet of the most exquisite sort as the two spacecraft pirouetted around each other, astronauts chattering on the radio and holding up signs for the cameras to see.

After 5 hours 18 minutes they backed away and Gemini 6A came home, having achieved one of the most dramatic missions to date. Gemini 7 followed behind just one day later – any more delays to Gemini 6A and the rendezvous would not have been possible. But Gemini 6A had proved that on-board navigation and careful maneuvers could bring two spacecraft together, while Gemini 7 proved two men could live in space long enough to get to the moon and back. An alarm call from space!

With Gemini 6 having been unable to chase down its

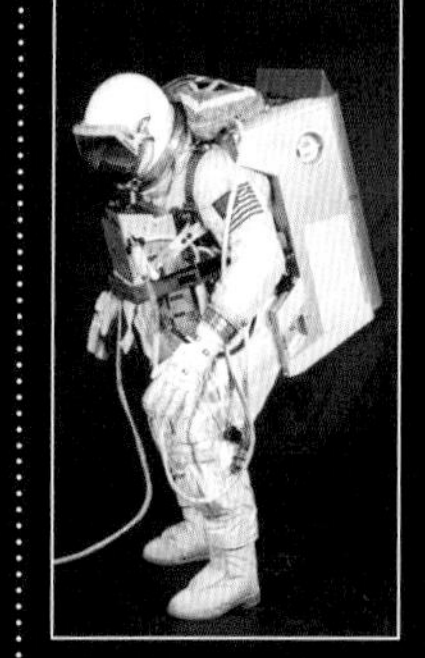

Dave Scott had planned to use a large backpack to test new ways of maneuvering around outside during EVA. A failed thruster on Gemini 8 caused an early return to earth and prevented that.

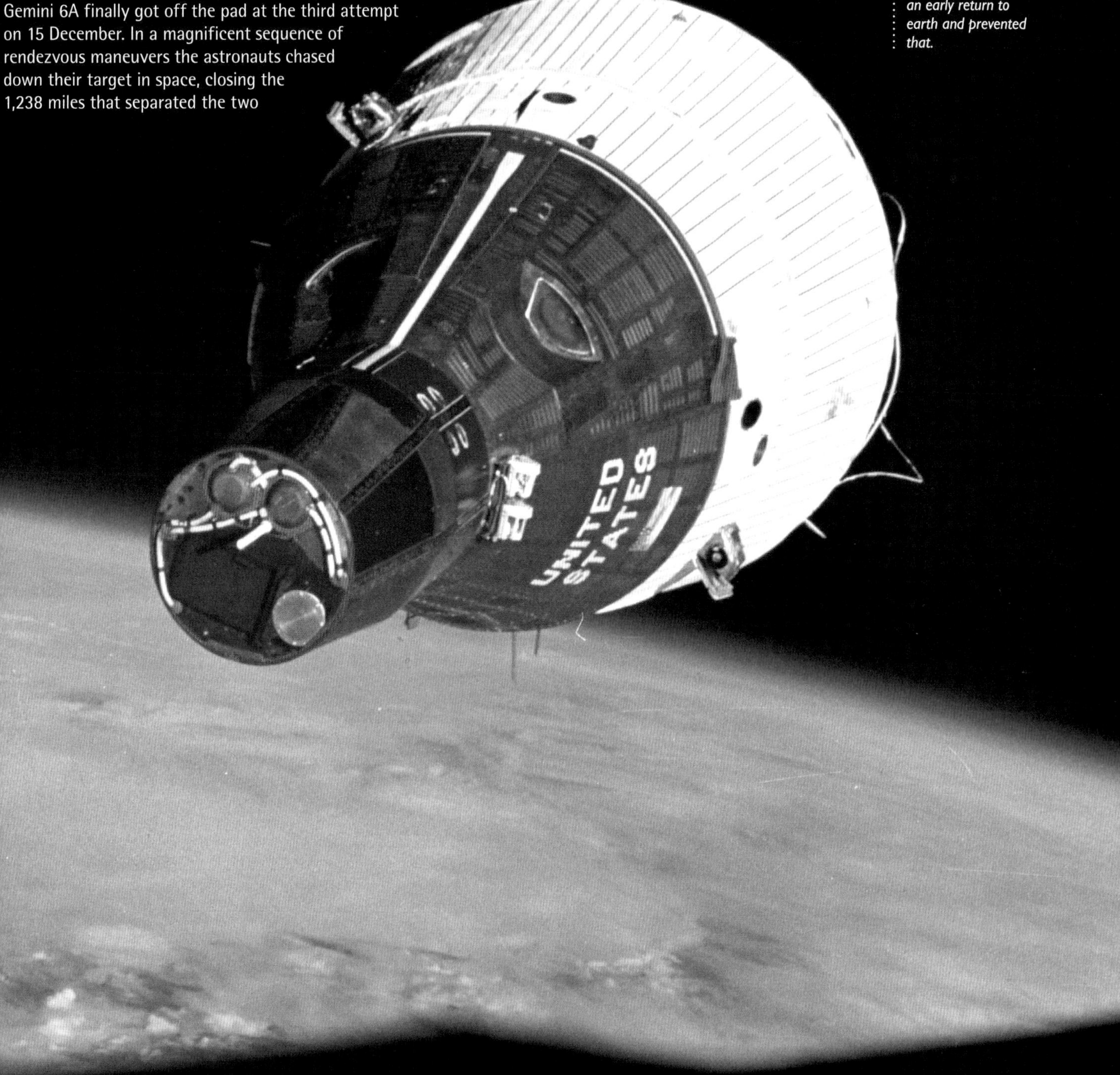

Navy divers drop to the sea to recover Gemini astronauts and place a flotation collar around the spacecraft.

The Gemini spacecraft is hoisted aboard the carrier where it is examined for possible damage during re-entry.

Agena in October 1965, the New Year brought hope that the last primary objective – the docking of two vehicles in space – would finally be accomplished. Having achieved rendezvous with Gemini 7, Gemini 6A had demonstrated there was little cause for concern over the last link in that chain: the direct linkup of two spacecraft in orbit.

Rookie pilots Neil Armstrong and Dave Scott thundered into space aboard Gemini 8 on 16 March, 1966. Before that, an Atlas had put their Agena target vehicle on orbit some 185 miles above the earth. Repeating the chase maneuvers adopted by the crew of Gemini 6A, Armstrong and Scott were alongside the Agena just 5 hours 54 minutes after their launch. Maneuvering round to the front of the Agena they slipped the nose of their spacecraft into a specially fitted docking cone and the two spacecraft formed a rigid structure, but not for long.

About 27 minutes after docking, with the spacecraft now out of communication with the ground, a stuck thruster began to spin the docked combination around, faster and faster until the crew was near to blacking out from the acceleration. Believing it to be a problem with the Agena, they undocked only for it to speed up until the spacecraft was spinning at almost 60

Milestones

1966

16 MARCH:
GEMINI 8 IS LAUNCHED AND DOCKS WITH AGENA BUT RETURNS EARLY

3 JUNE:
GEMINI 9 IS LAUNCHED BUT PILOT GENE CERNAN HAS TO ABORT A SPACE WALK WHEN HIS VISOR FOGS

18 JULY:
GEMINI 10 ACHIEVES A SUCCESSFUL DOCKING AND IS BOOSTED TO A RECORD 474 MILES ABOVE EARTH

12 SEPTEMBER:
GEMINI 11 DOCKS WITH ANOTHER AGENA AND IS BOOSTED TO 853 MILES, STILL A RECORD FOR A MANNED EARTH ORBITING SPACECRAFT

11 NOVEMBER:
GEMINI 12 IS LAUNCHED AND BUZZ ALDRIN CONDUCTS HIGHLY SUCCESSFUL SPACE WALKS

revolutions a minute. Only then did they identify the problem, quickly isolating the electrical circuit for that thruster and gradually regaining control with the other thrusters. But in getting Gemini back in a stable attitude, much of the thruster fuel that should have been used to control attitude during re-entry had been used up. Nothing for it but to return to earth as quickly as possible and that meant coming down in the Pacific less than eleven hours after launch.

Walking round the earth

The last four Gemini flights were in some respects the hardest of all – proving that astronauts could do useful work in space. The problem was that unless the weightless astronaut is adequately tethered to his work station, he drifts around – in trying to turn a screwdriver he merely turns himself around the screwdriver, which stays in one position! On earth, gravity keeps our feet firmly on the ground but in space tethers and restraints are needed to remain in one place. Any movement will cause the astronaut to drift away.

Launched in June 1966, Gemini 9's spacewalking astronaut 'Gene' Cernan had great difficulty trying to stay put while attempting to put on a backpack carried in the spacecraft adapter section. With a heart rate of more than

Cooper and Conrad arrive back on the deck of the US carrier Lake Champlain.

An Atlas rocket lifts the Agena target vehicle off the pad and into orbit. Six were launched but two (those for Gemini's 6 and 9) failed to reach orbit.

160 beats per minute Cernan was ordered back inside by command pilot Tom Stafford, after two hours struggling against aching muscles and a fogged up visor.

In July 1966, John Young and Mike Collins successfully rendezvoused and docked with another Agena target vehicle and using its rocket motor pushed them to a record height of 474 miles above the earth. They chased after the dormant Gemini 8 Agena and rendezvoused with it, proving that spacecraft could carry out complex and demanding maneuvers, practicing on-board navigation in a manner totally impossible with Mercury and Vostok.

Gemini 11 was equally successful in September that year when Conrad and Bean docked again with another Agena and, using its rocket engine, soared to yet another record height of 853 miles – a record that stands today as the highest altitude reached by a manned spacecraft in earth orbit. But it was left to 'Buzz' Aldrin on Gemini 12 to crack the problem with EVA. Aldrin used a water tank and divers' weights in a body belt to give himself 'neutral buoyancy, the closest simulation of weightlessness, so that he could help technicians devise appropriate tethers and restraints. Then he went into space and proved that it worked!

On the final Gemini mission, the last remaining problem had been solved and all the objectives had been met. It had taken just two years to propel America beyond the achievements of the Russians and as Gemini 12 came home in November 1966, Apollo 1 was being prepared on a Saturn launch pad at Cape Canaveral for its first manned flight three months hence. Nobody could know that a fatal flaw deep inside that spacecraft would bring disaster, threatening the goal set by President Kennedy less than six years earlier.

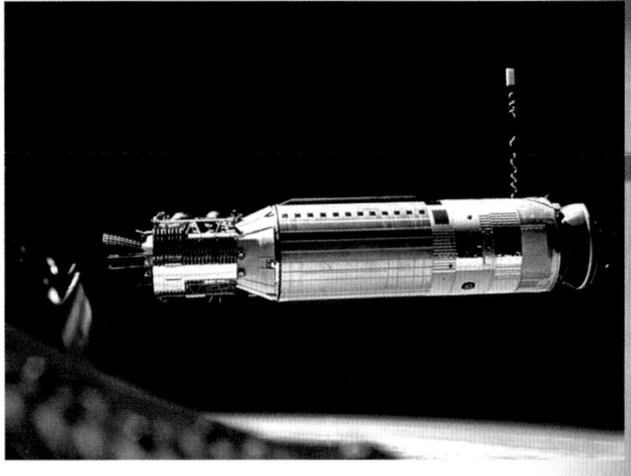

Gemini 8's Agena target vehicle shortly before docking.

BELOW • *Gemini 8 returns to a premature splashdown.*

BELOW RIGHT • *Conrad and Bean slam into the Atlantic Ocean after achieving what is still the highest altitude record of an earth orbit mission, ending the flight of Gemini 11.*

“ON THE FINAL GEMINI MISSION, THE LAST REMAINING PROBLEM HAD BEEN SOLVED AND ALL THE OBJECTIVES HAD BEEN MET. IT HAD TAKEN JUST TWO YEARS TO PROPEL AMERICA BEYOND THE ACHIEVEMENTS OF THE RUSSIANS...”

Buzz Aldrin gets it right on a spacewalk during Gemini 12.

RIGHT: *About to enter their spacecraft in an altitude chamber, the Apollo I crew engage in one of many simulations and tests prior to launch.*

BELOW LEFT: *The Apollo I spacecraft contained many flaws and its poor quality brought concerns at NASA but too late to prevent a disaster.*

BELOW MIDDLE: *Training to get out of an Apollo spacecraft after splashdown the crew takes to inflatable dinghies. Grissom (left) with White (foreground) and Chaffee find humour in the moment.*

BELOW RIGHT: *The crew for the first manned Apollo flight was selected on March 19, 1966, comprising (left to right) Grissom, White and Chaffee.*

A FATEFUL YEAR

After a run of 16 successful manned flights NASA was poised to fly Apollo – then disaster struck, and for the Russians too.

NASA had been racing to catch up with the Russians and now Apollo was being prepared for its first flight, a mission planned for February 1967. It was to be launched by a Saturn I from Launch Complex 34 at Cape Canaveral and veteran astronaut Virgil I 'Gus' Grissom had been selected to command a 10-day mission along with Gemini space-walker Ed White and rookie Roger Chaffee.

The spacecraft was of a Block I design, devoid of all the advanced technologies of the Block II model needed to fly all the way out to the moon and back. Fast-tracked to get it into space before the more sophisticated Block II, the mission was designated AS-204, after the rocket that would launch it, but everyone referred to it as Apollo 1. It would fly a routine earth-orbit mission checking out all the many systems and subsystems common to both models.

However, there had already been trouble with quality control at the manufacturer North American Aviation, and NASA had put in place troubleshooters to sort things out. Despite this, the astronauts knew that the ship was not ready to fly. There were too many bugs in the system and in simulations of the mission Grissom dubbed the spacecraft 'a lemon' and railed against persistent problems.

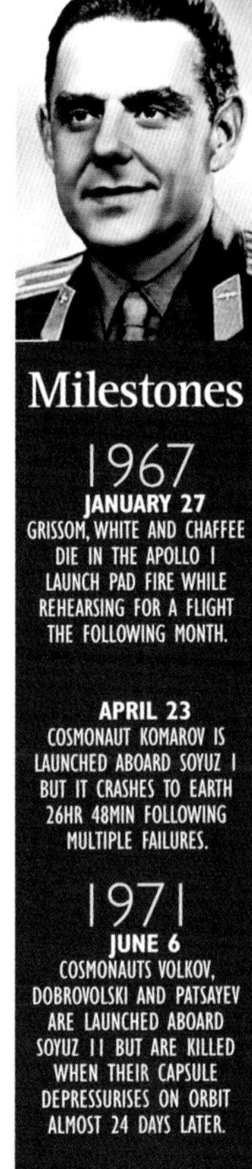

Milestones

1967

JANUARY 27
GRISSOM, WHITE AND CHAFFEE DIE IN THE APOLLO I LAUNCH PAD FIRE WHILE REHEARSING FOR A FLIGHT THE FOLLOWING MONTH.

APRIL 23
COSMONAUT KOMAROV IS LAUNCHED ABOARD SOYUZ I BUT IT CRASHES TO EARTH 26HR 48MIN FOLLOWING MULTIPLE FAILURES.

1971

JUNE 6
COSMONAUTS VOLKOV, DOBROVOLSKI AND PATSAYEV ARE LAUNCHED ABOARD SOYUZ II BUT ARE KILLED WHEN THEIR CAPSULE DEPRESSURISES ON ORBIT ALMOST 24 DAYS LATER.

Came the day

On January 27, 1967, Grissom, White and Chaffee entered their Apollo spacecraft on top of the launch vehicle at pad 34 and began a lengthy simulation of pre-flight activities. It was a routine test and the rocket was not fuelled. The crew had been sealed in their capsule for almost four hours when suddenly, at 6.30pm local time, an electrical spike from somewhere inside the capsule sent a spark to ignite inflammable materials. A member of the crew called out, 'fire in the spacecraft!' and within 20 seconds they were unconscious from asphyxiation as toxic fumes filled their suits. When the hatches were opened they were already dead.

Reaction

A long investigation followed the fire and NASA testified before Congressional committees horrified to learn that the spacecraft was highly vulnerable to fire and that electrical shorts were highly likely in what many termed a flawed design. Over the next year the spacecraft was completely refurbished with non-inflammable materials. But nothing could erase the memory of that fateful day when three astronauts lost their lives

Soyuz flies – and falls to earth

Russia had not launched a manned spacecraft for two years but now its new Soyuz spacecraft was ready and on April 23, 1967, it was launched into space carrying Valdimir Komarov, the first Russian to return to space. Form the moment it entered orbit things began to go wrong and a plan to launch a three-man crew aboard Soyuz 2 to join it the following day was abandoned. One of the Soyuz 1 fold-out solar panels failed to deploy and Komarov had to manually fire the retro-rocket for re-entry. During descent the main parachute failed and the back-up became tangled. Komarov died when Soyuz 1 slammed into the ground and burst into flames.

Two days later Komarov's remains were interred in the Kremlin wall. NASA astronauts Cooper and Borman contacted Moscow and offered to attend the funeral to express their condolences but they were told not to come, that it was 'an internal matter'. The death of their first cosmonaut brought as much grief to the Russian people as the loss of Grissom, White and Chaffee had to Americans. Now they were equals where no-one had wanted to be first. Four years later it happened again, when three cosmonauts lost their lives in Soyuz 11, June 1971, the last Russian space fatality.

TOP: *Like Apollo, Soyuz retained its external configuration as the spacecraft was thoroughly redesigned and made safe, but not before a second tragedy in 1971.*

TOP RIGHT: *Vladimir Komarov was selected for the first Soyuz but political pressure forced a flight no cosmonaut wanted to fly, in a spacecraft they all knew carried major problems.*

BELOW LEFT: *The fire created intense pressure inside the Apollo spacecraft, rupturing the base of the capsule. The head rests for the three couches are clearly visible where Grissom on the left was accompanied by White in the centre and Chaffee on the right couch.*

FAR LEFT: *Flames gushing out into the area around the capsule brought concern that heat could ignite the launch escape rocket, supported above by the lattice tower attached to the command module.*

APOLLO GETS READY

After the fire there was work to be done, making Apollo safe to fly.

By 1966 the Kennedy Space Center was beginning to look the part, directions on roadside boards approaching Cape Canaveral heralding the rocket site as 'Spaceport USA'. The giant Vehicle Assembly Building, 525 ft tall and more than 700 ft long, so big it could contain all the pyramids of Egypt, was ready for the giant Saturn V launch vehicles that would send Apollo to the moon. In this cavernous building would be assembled the biggest rockets ever flown into space, stacked components built all across the United States, flown in, brought by barge or trucked by road to this most symbolic of all places from where humans would depart planet earth for the first time.

Rollout

The Saturn V test vehicle, designated SA-500F, the F standing for facilities checkout vehicle, had been rolled to the launch pad on May 25, 1966, five years to the day after Kennedy announced the moon goal for Apollo. It tried out all the elements of Launch Complex 39 that would be used to take operational Saturns to the pad, two of which were more than three miles away

FAR LEFT • A full scale mock-up of the Saturn V rolls out to the launch pad, May 25, 1966. Left: The Apollo command module is recovered after the first Saturn V flight test.

BELOW, LEFT • The Launch Control Center housed 450 people working to get Saturn off the pad. Below and centre: The command module in assembly at North American Aviation.

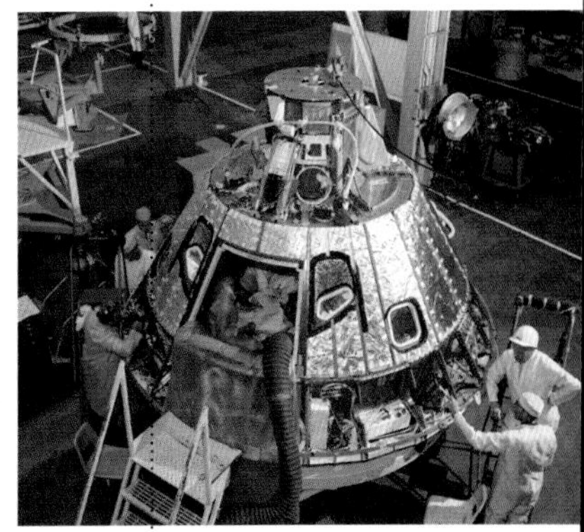

across reclaimed land. The first Saturn V was launched on November 9, 1967, on a mission designated Apollo 4. It carried a spacecraft similar to that which astronauts had been preparing to fly when fire swept through their capsule. The mission of Apollo 4 was to test the heat shield on the command module and to do that the capsule was fired to a maximum distance of 11,242 miles above the earth. Plunging back down into the atmosphere it successfully rode the roller-coaster path descending astronauts would fly coming back from the moon. Apollo 4's command module splashed down in the Pacific 8 hours 37 minutes after launch.

Built like a Swiss watch

Nothing in the Apollo programme demanded more attention than the safety of its astronauts and the redesign of the Apollo spacecraft after the fire occupied several thousand engineers and technicians for more than a year. The spacecraft itself was a pressure-can filled with an oxygen-nitrogen atmosphere on the pad, gradually replaced with pure oxygen as it entered space. Around the outside of the conical command module were equipment bays where the systems and subsystems were housed. Surrounding it all was a heat shield made of phenolic epoxy resin injected as a jelly-like substance into a stainless-steel honeycomb matrix, where it would set hard. During descent from orbit, or returning from the moon, the heat shield would char away, taking away the heat caused by friction with the atmosphere. This process of ablation wore away the surface of the heat shield, leaving about half the thickness it had when it was installed around the spacecraft.

OPPOSITE PAGE, BELOW • The Saturn V was 363ft tall and weight more than 2,500 tons at liftoff.

Milestones

1966

25 MAY:
A FULL-SIZE REPLICA, SATURN V SA-500F IS ROLLED OUT TO THE PAD TO CHECK OUT THE FACILITIES ON LAUNCH COMPLEX 39.

1967

26 AUGUST:
APOLLO 4, SATURN V AS-501, IS ROLLED FROM THE VAB OUT TO LAUNCH COMPLEX 39A.

9 NOVEMBER:
APOLLO 4 IS LAUNCHED CARRYING AN UNMANNED APOLLO SPACECRAFT ON A HIGH ALTITUDE TEST OF SPACECRAFT AND HEAT SHIELD.

1968

4 APRIL:
LAUNCHED BY SATURN V AS-502, APOLLO 6 CARRIES AN APOLLO SPACECRAFT ON A TEST SHOT TO ORBIT, THE SPACECRAFT RETURNING TO A SPLASHDOWN IN THE PACIFIC OCEAN.

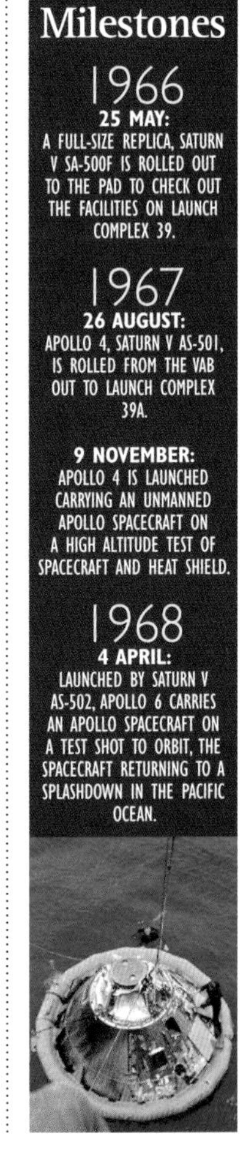

A Saturn IB launches the first manned Apollo space mission in October 1967.

ABOVE RIGHT • Schirra (centre), Eisele (left) and Cunningham, the crew of Apollo 7

RIGHT • Technicians help the astronauts ease onto the couches inside the command module.

BELOW RIGHT • Parody of a US TV show, the crew hold up a notice inviting letters!

MOONWALK USA

A time to test

The tragic fire of January 27, 1967, that took the lives of three NASA astronauts did little to deter the spirit of those working to put an American on the moon by the end of the decade. The spacecraft itself was modified to prevent a fire taking hold, even the flight plans were made from non-inflammable paper and fires were deliberately created in mockups duplicating a real spacecraft to show that, once started, flames would be automatically extinguished. So it was with a sense of renewed optimism that Apollo astronauts got ready to test fly their new spacecraft in earth orbit on a 10-day mission called Apollo 7. Before that, extensive rehearsals and simulations tested man and machine to the limit while scientists and engineers planned in detail one of the most rigorous checkouts of any spacecraft in orbit.

Apollo flies

Veteran astronaut 'Wally' Schirra, the only astronaut to fly Mercury, Gemini and Apollo, commanded a mission that included rookie crewmembers Don Eisele and Walter Cunningham. Launched on October 11, 1967, Apollo 7 rode the most powerful rocket used by astronauts thus far, the Saturn IB. With a thrust of 1.5 million lb, it was almost four times more powerful than the Titan 2 used for Gemini flights two years earlier, but it did its job well and placed the 16 ton spacecraft in earth orbit. For the prescribed 10 days it orbited the earth checking and testing every element to prove the spacecraft design could be relied upon to carry astronauts all the way out to the moon and back.

But there was a very different kind of test as well when the snappy commander developed a cold in space and gave his crewmates a hard time, brusquely pushing aside demands from Mission Control for TV pictures to show on national television. The public relations people at NASA wanted to show off their brand new spacecraft to the American taxpayers, but Schirra was not about to be told what to do – and he did the broadcasts when he wanted to, not when Houston told him! Nevertheless, the

Milestones

1968

7 AUGUST: NASA MANNED FLIGHT BOSS GEORGE MUELLER PROPOSES SENDING THE APOLLO 8 MISSION AROUND THE MOON IN DECEMBER, SUBJECT TO SUCCESS WITH APOLLO 7.

14 SEPTEMBER: THE RUSSIANS WERE BELIEVED TO BE PREPARING FOR A MANNED FLIGHT AROUND THE MOON BEFORE THE END OF THE YEAR.

11 OCTOBER: APOLLO 7 IS LAUNCHED ON A 10-DAY PROVING FLIGHT FOR THE NEW SPACECRAFT TO TEST ALL ITS SYSTEMS

22 OCTOBER: SCHIRRA, EISELE AND CUNNINGHAM RETURN TO EARTH AFTER NEARLY 11 DAYS IN SPACE ABOARD APOLLO 7

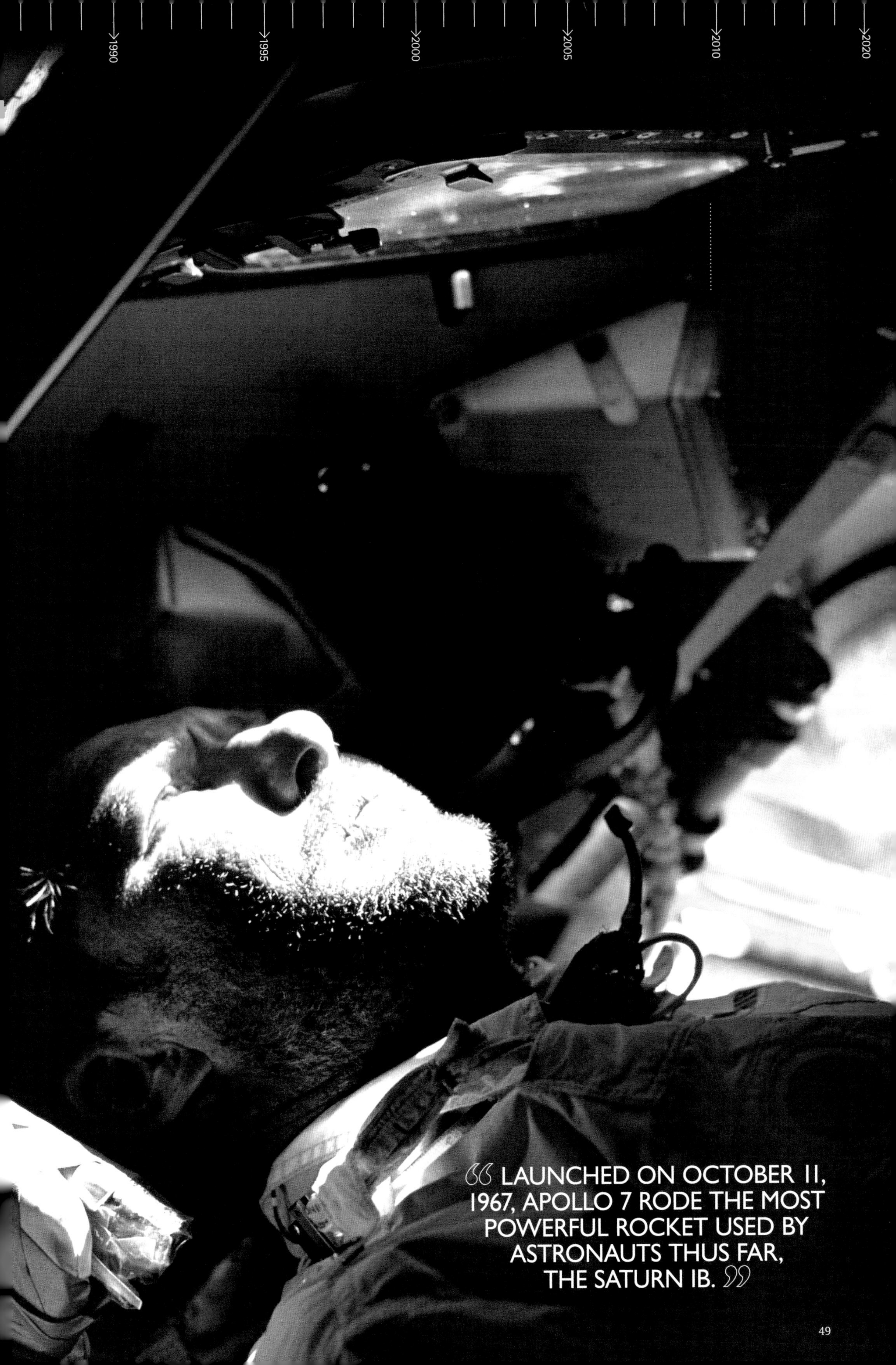

“LAUNCHED ON OCTOBER 11, 1967, APOLLO 7 RODE THE MOST POWERFUL ROCKET USED BY ASTRONAUTS THUS FAR, THE SATURN IB.”

Petal-like doors that had supported the Apollo spacecraft on top of the second stage of the Saturn IB.

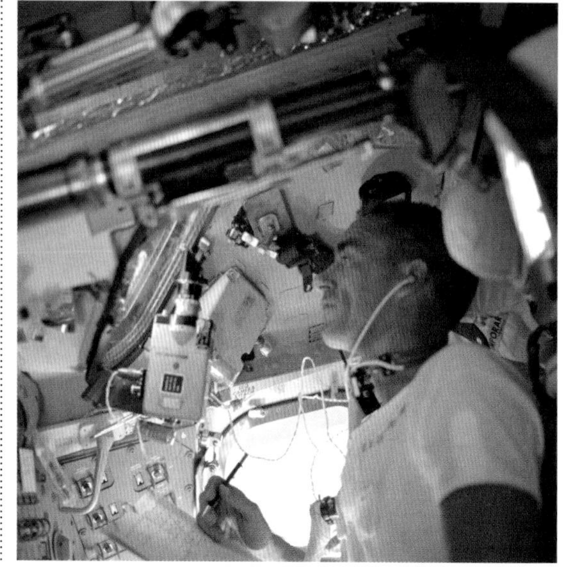

Apollo 7 crewmembers take readings during their long mission.

first live TV pictures from an American spacecraft went down well and would become a routine part of space flight from then on.

Testing the power

Because Apollo was a completely new spacecraft a vital part of the 10-day flight was to check out the guidance and navigation tasks essential to back up instructions from the ground. On missions to the moon, where a breakdown in communications with earth could leave crewmembers without vital data voiced up from earth, the crew would have to take star measurements and navigate themselves back home, so those activities were tested in earth orbit – close to home!

Another test called for the crew to fire the powerful Service Propulsion System, a rocket motor in the service module to which the cone-shaped command module was attached. On moon missions it would be used to put the combined Apollo and Lunar Module in moon orbit, and to fire again to bring Apollo back home when the two moonwalkers rejoined the lone astronaut in the command module. With a thrust of 20,000lb, it was a powerful rocket motor and it performed well on eight firings to change orbit, the final one lasting 12 seconds to slow the spacecraft and bring the crew home at the end of the mission.

When Apollo 7 splashed down in the Atlantic 10 days and 19 minutes after liftoff, it had demonstrated the success achieved by NASA and its contractors in building a much safer spacecraft. It also demonstrated new technology first tested on Gemini missions, fuel cells for electrical power that also produced water as a by-product. The way was now clear for a bold step forward.

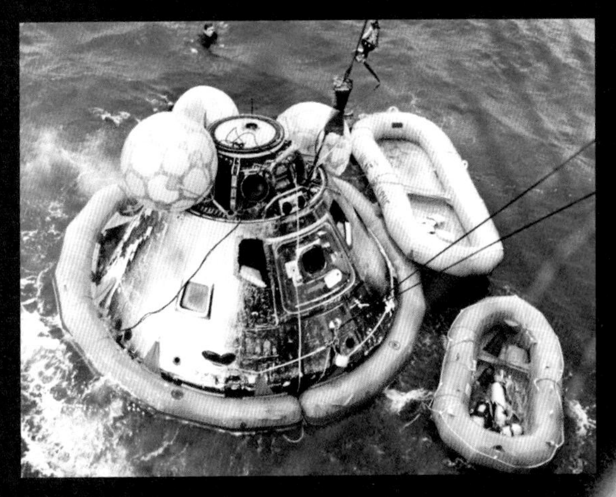

The Apollo command module had three 'golf-ball' flotation bags for righting the spacecraft should it tip over on splashdown.

" WHEN APOLLO 7 SPLASHED DOWN IN THE ATLANTIC 10 DAYS AND 19 MINUTES AFTER LIFTOFF, IT HAD DEMONSTRATED THE SUCCESS ACHIEVED BY NASA AND ITS CONTRACTORS IN BUILDING A MUCH SAFER SPACECRAFT "

(From left) Lovell, Anders and Borman at the stairs of the spacecraft simulator.

RIGHT • *The launch of Apollo 8, the first manned flight of a Saturn V.*

BELOW: *Crowds watch the launch from a special VIP stand at the Kennedy Space Center*

To boldly go...!

When NASA planned its Apollo missions it wanted to follow the first manned flight with a full test of both the Apollo spacecraft and the Lunar Module on the giant Saturn V specially designed and built for blasting the 45 ton combination of mother-ship and moon lander to the moon. That first flight, known as a 'shakedown test', would be a full dress rehearsal in earth orbit, before sending a similar combination all the way to the moon and back. But the Lunar Module was late and technical difficulties with getting it ready for flight were threatening to delay it beyond the planned launch date of December 1968. The possibility existed of delaying the first flight of the combined Apollo/Lunar Module to the next mission, Apollo 9, and using the Apollo 8 launch slot for sending just the Apollo spacecraft on a daring mission all the way to the moon and back, spending ten orbits over the lunar surface before returning home.

It was daring indeed and not everyone at NASA thought it wise to risk a new spacecraft, flown only once in earth orbit, on such a risky venture. It would take three days to get to the moon and three days to get back, with almost a full earth day in moon orbit. There was no chance of a fast return to earth and if anything went wrong with the spacecraft it could not only threaten the lives of the crew but also the possibility of landing on the moon before the end of the decade – and there was only a year left to go!

With a successful Apollo 7 test flight under their belt, NASA managers finally decided to go for a lunar orbit mission with Apollo 8 in December, deferring the first combined test of Apollo and Lunar module to Apollo 9 early in 1969. Astronauts Borman, Lovell and Anders would be the first humans to leave the gravity field of their home planet.

'The big Daddy!'

When a rocket engineer who had been working boosters at the Cape for more than a decade saw the first rollout of a Saturn V mock-up in 1966, he was sure that this was 'the big Daddy of them all!' And it was. With a height of 363ft, the Saturn V was five times more powerful than the Saturn IB that lifted Apollo 7 into space and it was more than 16 times as powerful as the Titan 2 that launched Gemini. To seasoned space workers across the Cape, and at Mission Control in Houston, the launch of Apollo 8 on the morning of December 21, 1968, was one

Milestones

1968

11 NOVEMBER: NASA'S ACTING ADMINISTRATOR THOMAS PAINE APPROVED THE PLAN TO FLY APOLLO 8 INTO MOON ORBIT

23 NOVEMBER: SOVIET SPACE OFFICIALS ANNOUNCED THEY WERE READY TO SEND A MANNED SPACECRAFT AROUND THE MOON AND THAT SIX COSMONAUTS HAD PRESSED FOR THE POLITBURO TO APPROVE SUCH A MISSION. IT DID NOT.

21 DECEMBER: BORMAN, LOVELL AND ANDERS ARE THE FIRST ASTRONAUTS TO RIDE THE SATURN V AND THE FIRST TO BE PROPELLED TOWARD THE MOON

24 DECEMBER: APOLLO 8 ENTERS LUNAR ORBIT FOR 10 REVOLUTIONS OF THE MOON

25 DECEMBER: BORMAN, LOVELL AND ANDERS HEAD BACK TO EARTH 1968, 27 DECEMBER: THE APOLLO 8 SPLASH DOWN IN THE PACIFIC OCEAN

Even at a distance of more than three miles, the thunder of the Saturn V was a noise so loud it was almost unbearable, a power equal to almost 300 million horsepower lifting a 3,000 ton rocket as tall as St Paul's Cathedral free of earth's gravity and on course for the moon at 25 times the speed of sound. But first, a couple of earth orbits before firing out toward the moon.

As the third stage lit up and pushed Apollo free of earth's gravity at the start of its three-day journey, it was a moment dreamed of for centuries – humans from earth were on their way to earth's nearest neighbour in space. On the way the crew sent back fuzzy TV pictures of their home planet claiming it to be a 'grand oasis in the vast loneliness of space'. The time went fast for those watching and waiting on earth. As the spacecraft slipped behind the left side of the moon and disappeared around to the far side, controllers on the ground could only wait in silence trusting that the big service propulsion system engine on Apollo had fired on time to place them in orbit.

Right on time they reappeared around the right side of the moon, for the first time in orbit about another body in space. It was Christmas Eve and almost the entire world, it seemed, was tuned in to radio and TV sets, anxiously hanging on every word from 240,000 miles away across the void. Slowly, in an almost monotone voice, the crew began to read from the first chapter in the book of Genesis: 'In the beginning, God created the heaven and the earth, and the earth was without form and void, and darkness was upon the face of the deep...', repeating verses that underpinned all the world's major religions throughout human history, bringing a special poignancy to this Christmas Eve, as an estimated one billion people around the world listened as one.

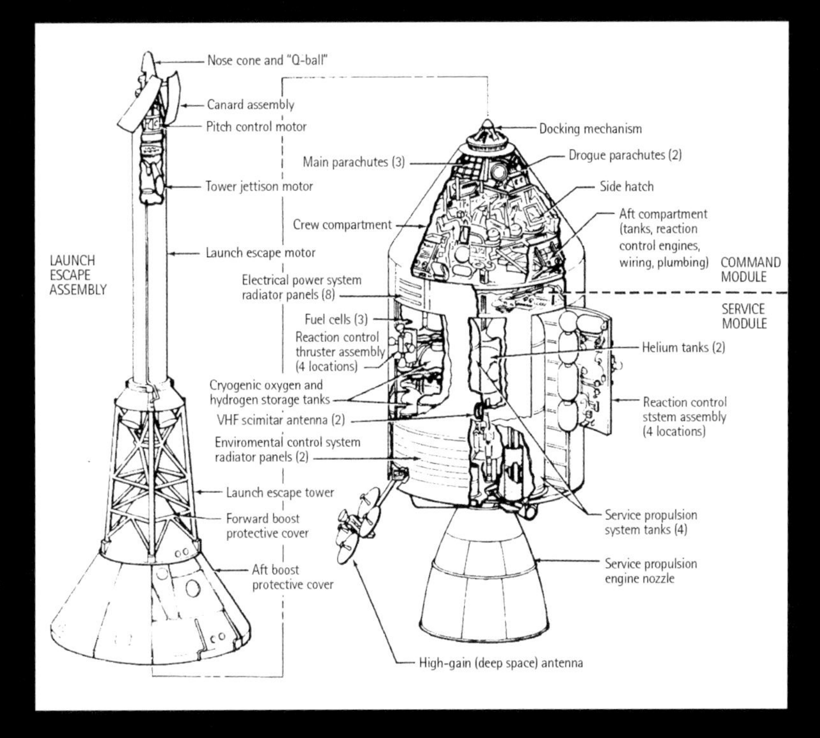

The crew occupied the command module, situated above the cylindrical service module which had all the equipment necessary for up to 14 days in space.

...There is a Santa Claus!

After 20 hours in lunar orbit, the main engine was fired again to push Apollo 8 out of lunar orbit and on its way back to earth. On appearing around the right side of the moon, the voice of Jim Lovell reported to earth: 'Please be informed, there is a Santa Claus!' They were on their way home. The spacecraft splashed down in the Pacific Ocean on December 27 and the crew began a period of intense debrief followed by a world tour. But there was no rest for scientists and engineers at NASA, as preparations swung into action for the last dress rehearsals before the first attempt at a moon landing.

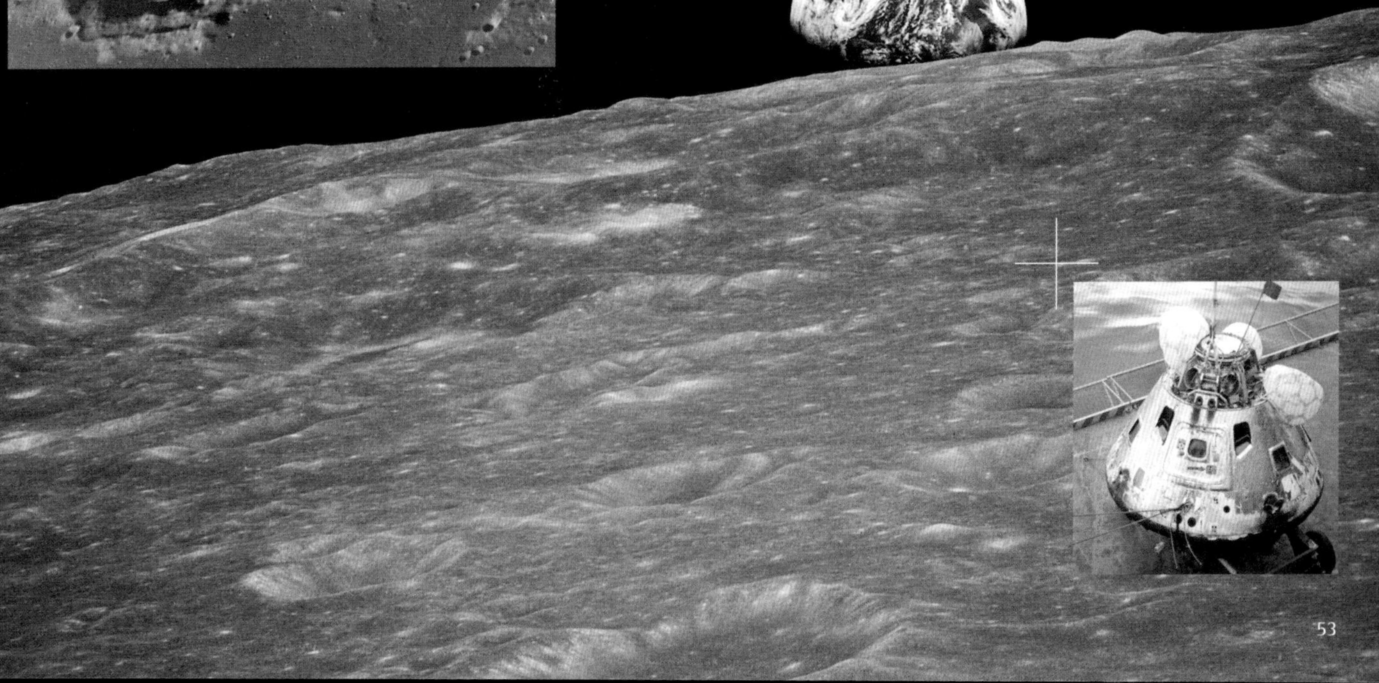

Earthrise from lunar orbit and (left) the craters photographed from close up for the first time.

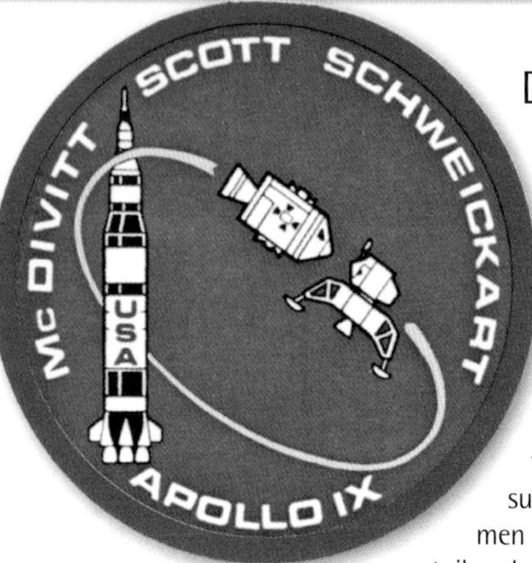

Dress rehearsal

When the final year of the decade began in January 1969, only two more flights remained until the attempt to land on the lunar surface. Across the United States and at tracking stations around the world, technicians and engineers were gearing up to put into place the final pieces of the giant Apollo jigsaw that would carry two men to the lunar surface. It had taken around 400,000 men and women eight years of non-stop toil and self-sacrifice to get to the point where everything was ready. But there was still the Lunar Module to test and that would be put through its paces during the flight of Apollo 9.

A year earlier, on January 22, 1968, a Saturn IB had carried a Lunar Module into earth orbit for a remotely controlled test of its basic systems. There had been problems on that flight, some of which had delayed the completion of a fully complete Lunar Module qualified for manned operations. But finally, all was ready and on March 3, 1969, the 16 ton Lunar Module named 'Spider' was launched with the 28 ton Apollo spacecraft called 'Gumdrop' on top of a Saturn V from Cape Canaveral. On board Apollo 9 were astronauts McDivitt, Scott and Schweickart.

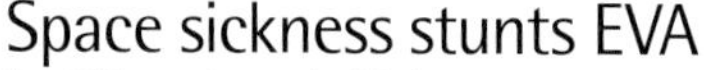

Space sickness stunts EVA

It afflicts about half the astronauts that go into space and it is particularly disabling but space sickness is every bit as real as motion sickness on earth and is largely triggered for the same reason – instability in the middle ear where the body senses disorientation. Rusty Schweickart got it and it compromised his space walk, the first EVA of the Apollo programme and the first since the last Gemini flight in November 1966. His job was to demonstrate how, in an emergency, an astronaut could get from one spacecraft to the other via an external route between hatches on the Lunar Module and the Apollo spacecraft. He managed to put on his suit and climb out of the Lunar Module but fearing he would vomit in his helmet the test was considered complete and he quickly got back in!

The flight lasted just over 10 days, during which the two spacecraft separated and the Lunar Module fired the engine that was designed to land the LM on the moon,

ABOVE RIGHT • *Apollo 9's crew of (left to right) McDivitt, Scott and Schweickart.*

RIGHT • *The Lunar module 'Spider' as viewed from the command module 'Gumdrop'*

BELOW RIGHT • *Dave Scott stands in the open hatch of the command module as he is photographed by Schweickart from the 'porch' of the Lunar Module.*

Milestones

1968

22 JANUARY:
SATURN IB AS-204, THE ROCKET THAT SHOULD HAVE LAUNCHED APOLLO, PUT THE FIRST LUNAR MODULE IN ORBIT ON AN UNMANNED TEST

1969

3 MARCH:
SATURN V AS-504 CARRIES THE FIRST MANNED LUNAR MODULE ON THE 10-DAY APOLLO 9 EARTH ORBIT SHAKEDOWN FLIGHT WITH ASTRONAUTS MCDIVITT, SCOTT AND SCHWEICKART

18 MAY:
AS-505 LAUNCHES APOLLO 10 FOR A FULL LUNAR LANDING REHEARSAL AROUND THE MOON

LEFT • (left to right) Cernan, Young and Stafford pose in front of their Saturn V.

and the ascent engine that would be used to get it off the surface again. Both systems came through with flying colours and when the flight ended it had cleared the way for the last full-dress rehearsal before the landing attempt.

The final run

The year 1969 was the year of the moon. NASA increased its flight preparations to get in at least five Saturn V flights in an attempt to make it down to the surface. With Apollo's 7, 8 and 9 a success, it was hoped that after one final full-scale rehearsal around the moon in May, Apollo 11 would target the surface during July. If not, if something went wrong on either of the next two missions to prevent that, Apollo's 12 and 13 would be available for another try.

Launched on May 18, 1969, Apollo 10 thundered away

and Cernan. First a checkout of all spacecraft systems in earth, then, on the second revolution, the third stage of Saturn V fired to blast its cargo away from the earth and toward the moon. Just under three days later it arrived in lunar orbit and over a period of little more than 61 hours around the moon demonstrated that the Lunar Module could operate safely as a separate spacecraft from Apollo and that all the communications, life support and propulsion systems could support a lunar landing.

The toughest cut of all was the simulated dive down toward the lunar surface, climbing back again little more than 8 miles above the moon! But it proved the system would work and the path was now clear for Apollo 11 to go all the way. When Apollo 10 returned home nothing now stood in the way of the first moon landing.

LEFT • The ascent stage of the Lunar Module comes back to link up with the command module 50 miles above the moon.

Man on the Moon!

The decision to send men to the moon in a race with the Russians had been taken in late May 1961. Less than eight years and two months later, in a race unlike anything ever mounted in history, that goal was accomplished with the almost perfect flight of Apollo 11. Launched on July 16, 1969, astronauts Armstrong, Collins and Aldrin were sent on their way by a Saturn V rocket, around a million people watching from the coast-hugging highways around Cape Canaveral and Cocoa Beach for miles across the coast of Florida. They came from around the world, more than 50,000 flying in to America to witness first hand the event that epitomised all of human ambition for centuries since the first people gazed in awe at the night sky.

TOP RIGHT • (left to right) Armstrong, Collins and Aldrin, the crew of Apollo 11.

ABOVE • Time to go!

Shortly before the launch, the gates of the Kennedy Space Center had been thrown open to admit a group of Civil Rights protesters complaining at the lack of attention to poverty and to proclaim the needs of black communities across America. As they stood with center director Kurt Debus, who had accompanied Wernher von Braun from Germany in 1945 to help build America's rocket programme, tears ran down their faces. They stood in silence, shaken by the thunder of the Saturn V, awe-struck at the sheer spectacle and sharing the moment.

Across America, hospitals reported a decline in death rates as the elderly clung on to life, determined to be witness to mankind's greatest adventure, and police stations in many US States reported a reduction in crime rates. Television sales reached an all-time high and around the world, again more than a billion people followed events hour by hour.

RIGHT • The earth, seen shortly after heading out for the moon.

BELOW RIGHT • Buzz Aldrin hops down the ladder to the lunar surface.

Milestones
1969
16 JULY:
APOLLO 11 IS LAUNCHED FROM THE KENNEDY SPACE CENTER
20 JULY:
CARRYING ARMSTRONG AND ALDRIN, LUNAR MODULE EAGLE LANDS ON THE MOON, MIKE COLLINS REMAINING IN ORBIT ABOARD APOLLO, NAMED COLUMBIA
24 JULY:
APOLLO 11 SPLASHES DOWN IN THE PACIFIC AND IS MET BY PRESIDENT NIXON

'...30 seconds...!'

The flight of Apollo 11 was textbook perfect, up to the point where the spacecraft separated in lunar orbit and started down toward the surface. Beginning from a height of less than 10 miles above the moon, the Lunar Module carrying Armstrong and Aldrin flew parallel to the surface, gradually losing height. They were flying feet first, lying on their backs. At around 24,000ft the Lunar Module called Eagle began to pitch forward and the landing site came into view. Except it was not quite where it should be. Eagle was a little long and to one side of the predicted ground track. Moreover, it became difficult to see the surface clearly. Dust was scattering and obscuring the view.

By prior arrangement, Mission Control would call out the time remaining before the fuel ran out, at which point they would have to abort. They would do that by firing up the engine on the Ascent Stage to get them back into orbit. In Mission Control flight director Gene Kranz ordered quiet. Only two voices broke the silence – Buzz Aldrin in Eagle calling out the readings and Charlie Duke in Mission Control, talking to the crew.

As the seconds ticked down and Duke called '30 seconds!' it seemed a long time before the word from Neil Armstrong that everyone wanted to hear: 'Houston, Tranquillity Base here, the Eagle has landed'. There was less than 20 seconds to spare. It was July 20, 1969 and less than seven hours later Armstrong put his bootprint on the dusty surface with the words 'That's one small step for a man, one giant leap for mankind.'

The Washington Post
'The Eagle Has Landed'—
Two Men Walk on the Moon

One of the greatest public events of the century in which a billion people around the world watched history's greatest day.

On the airless moon, bootprints will last for millions of years.

LIFE
ON THE MOON
Footprints and photographs by Neil Armstrong and Edwin Aldrin

Buzz Aldrin unpoacks equip,ment from the back of the Lunar Module.

Neil Armstrong and the Lunar Module reflected in the visor of Buzz Aldrin's helmet.

A new moon

The race had been won, the moon had been conquered and the lives of at least three men would change for ever. Standing outside looking at the moon, Werhner von Braun found a parallel with the evolutionary story of living things, believing this event to be as important as 'the day aquatic life crawled out on land for the first time.' And in Russia, their race lost, cosmonauts reflected on the words of a Russian mathematician Konstantin Tsiolkovsky who at the turn of the century had foretold the destiny of man in space: 'The earth is the cradle of the mind, but one cannot live for ever in a cradle.'

ABOVE: *A gold plated olive branch as a symbol to accompany the words on a plaque on a leg of the Lunar Module, 'We Came in Peace for all Mankind.'*

LEFT • *Setting ouit scientific experiments. The device in front is a seismometer to measure moonquakes.*

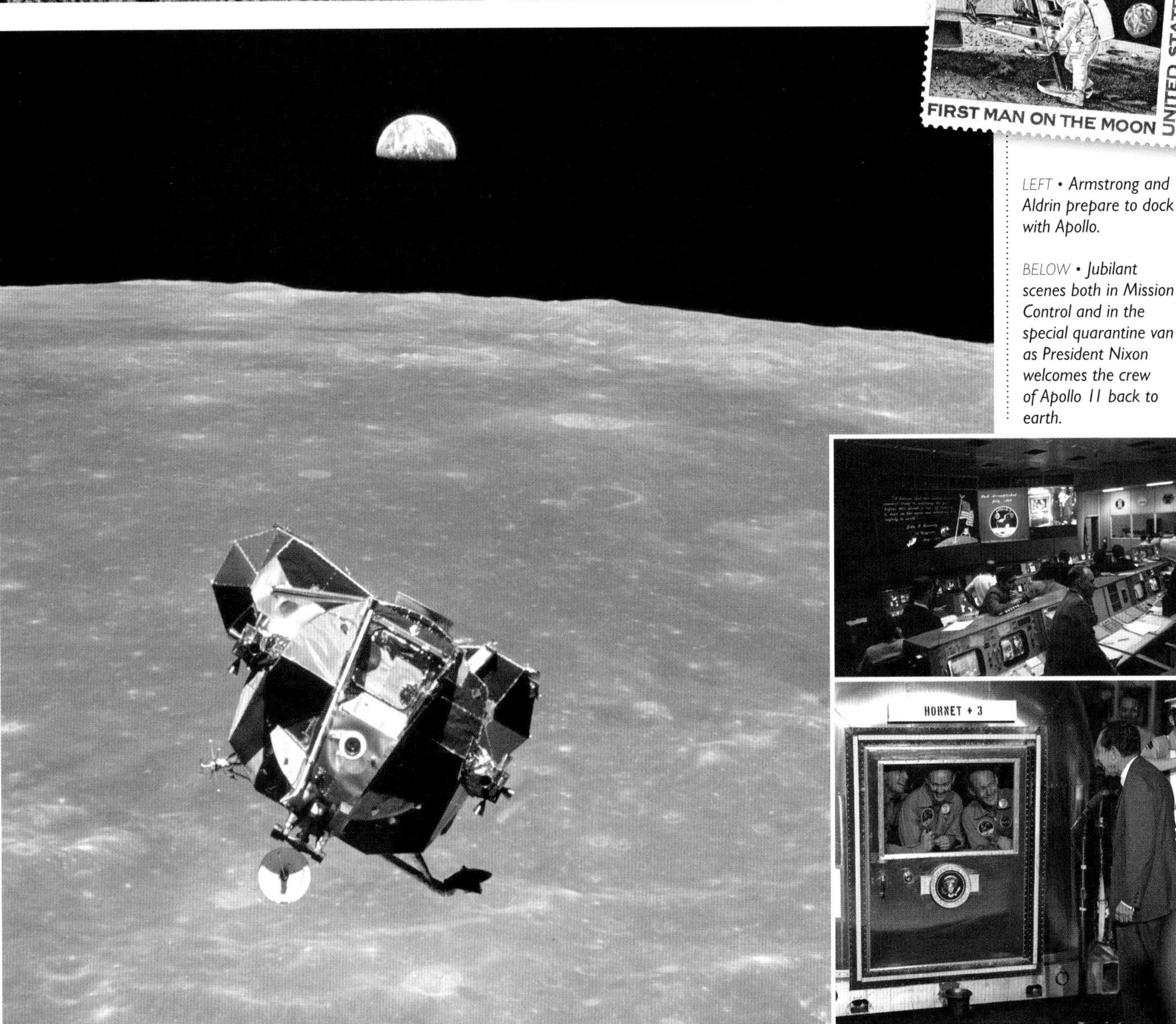

LEFT • *Armstrong and Aldrin prepare to dock with Apollo.*

BELOW • *Jubilant scenes both in Mission Control and in the special quarantine van as President Nixon welcomes the crew of Apollo 11 back to earth.*

RENDEZVOUS WITH A ROBOT

Precision parking

Getting to the moon was just the beginning. Doing useful scientific exploration was the main goal after the race – and that had been won when Armstrong and Aldrin put Eagle on the lunar surface. But Apollo 11 had landed more than four miles off target. If astronauts were ever to get into difficult and mountainous places to do useful work it was necessary to show that a Lunar Module could land with great precision at a particular spot on the moon. There was no better way to demonstrate that than by landing alongside a robot spacecraft sent to the moon several years earlier.

In trying to find out as much about the condition of the lunar surface before sending men, NASA landed five Surveyor spacecraft at different places. One of these, Surveyor 3, was sitting on a relatively smooth plain suitable for a Lunar Module to put down. But with the pressure off, the launch rate of a mission every two months was relaxed and not until November 1969, four months after Apollo 11, did Apollo 12 leave for the moon.

Apollo 12 launched into rain and an electrical strike from a nearby thunderstorm suddenly sapped power from Apollo's instruments. 'What a ride, what a ride! I'm not sure we didn't get hit by lightning,' cried an excited Pete Conrad as he and fellow astronauts Dick Gordon and Alan Bean hurried to switch everything back on while still hurtling to orbit. It was a hairy ride but they made it and quickly settled down to a three-day drift to their destination in the Ocean of Storms on the lunar surface, landing just a few hundred feet from Surveyor 3.

Snap-happy

Hopping down the ladder on the front leg of the Lunar Module Intrepid, the diminutive Conrad parodied

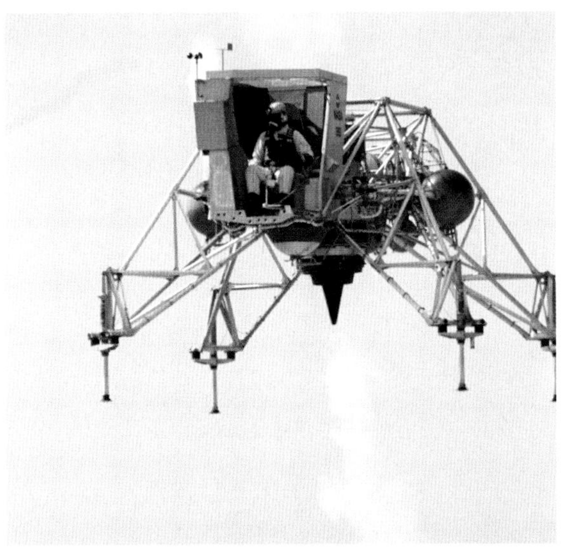

Far left: Preparations to fly the Lunar Module began with a workout on the Lunar Landing Training Vehicle, simulating the vertical lift of the Lunar Module.

Left: Famous for its naval aircraft, Grumman built the Lunar Module at its Bethpage facility.

Below: Apollo 12 astronauts deployed an erectable dish-shaped antenna for transmitting TV signals direct to earth.

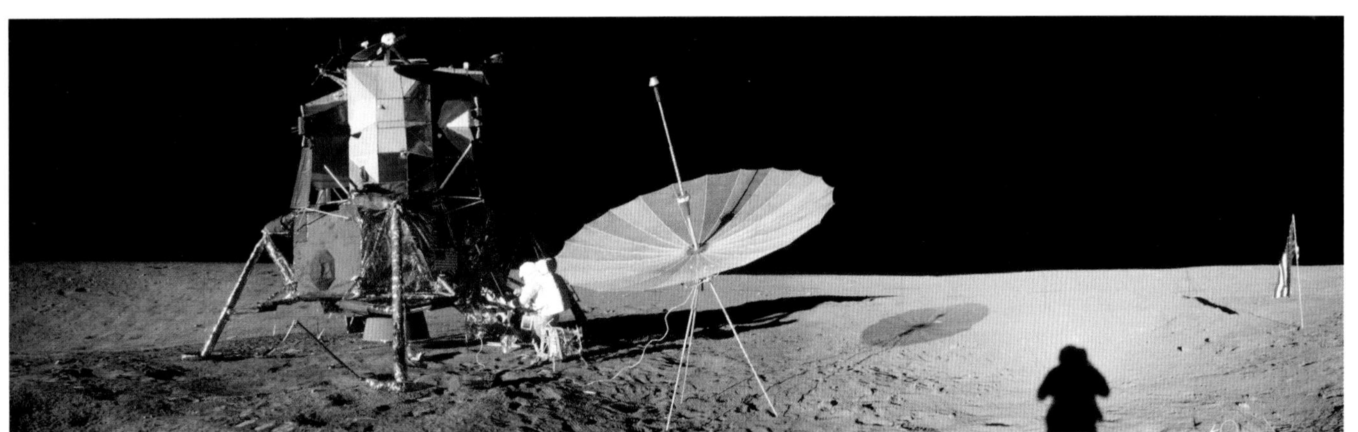

Armstrong's first words on the surface when he called out to earth, 'Whoopee, man that may have been a small step for Neil but it's a long one for me!' Eager to take pictures, and sending broadcasts back to earth, Conrad set up a new and improved TV camera but Alan Bean inadvertently pointed it at the sun and burned out the vidicon tube. Fuming at the lost publicity, NASA placated irate TV stations who now had only sound to convey activities from the lunar surface, which would, they said, have been spectacular. It didn't help!

There were two space walks, the first laying out an array of science instruments powered by a radioactive power module. The second EVA took Conrad and Bean around the crater within which sat Surveyor 3, from which they snipped the TV camera off the robot lander. What very few back on earth knew was that Playboy centerfolds had been cut out, reduced in size and pasted in to their checklists strapped to their suit cuffs! Always up for a practical joke, Conrad and Bean smuggled a timing device aboard the spacecraft with the idea of taking a picture of themselves, claiming there was a third person with them on the surface - but they lost the timer and the opportunity for mischief.

Bugs in the works

On examining the Surveyor 3 TV camera back in their laboratories on each, scientists found microscopic bacteria inadvertently carried from earth had been growing on the lunar surface. The source was traced to a technician wearing a sterile suit,who had sneezed while the robot was being prepared for launch. The first indication that living things can survive in the hostile environment of space.

Left: (left to right) Conrad, Gordon and Bean pose in front of a Lunar Module mock-up.

Below: samples lifted from the surface will be pored over by scientists around the world, presenting a very different chemistry to earth rocks.

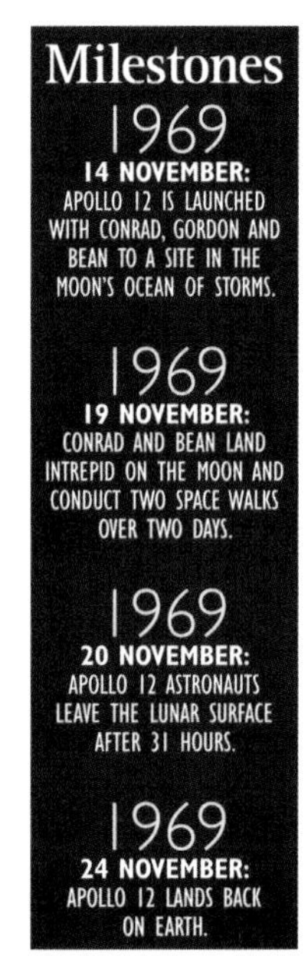

Milestones
1969
14 NOVEMBER:
APOLLO 12 IS LAUNCHED WITH CONRAD, GORDON AND BEAN TO A SITE IN THE MOON'S OCEAN OF STORMS.
1969
19 NOVEMBER:
CONRAD AND BEAN LAND INTREPID ON THE MOON AND CONDUCT TWO SPACE WALKS OVER TWO DAYS.
1969
20 NOVEMBER:
APOLLO 12 ASTRONAUTS LEAVE THE LUNAR SURFACE AFTER 31 HOURS.
1969
24 NOVEMBER:
APOLLO 12 LANDS BACK ON EARTH.

“HOUSTON – WE HAVE A PROBLEM...”

The luck of the draw

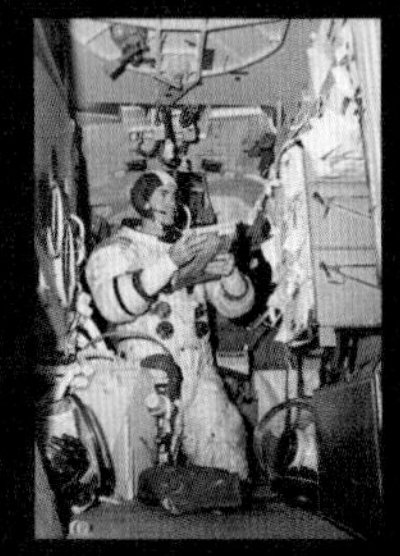

BELOW • A tired crew grab a few 'souvenir' pictures as the crippled Apollo spacecraft zips round the far side of the moon. Inset: (left to right) Lovell, Swigert and Haise, the crew of Apollo 13.

Along with Jim Lovell and Fred Haise, Ken Mattingly had been picked to fly NASA's third flight to the moon but less than a week before the launch of Apollo 13, he was exposed to German measles. Back-up astronaut Jack Swigert would take Mattingly's seat. Right on time at 13.13 hours on April 11, 1970, Apollo 13 took off and almost immediately ran into trouble when one of the five rocket engines on the Saturn V's third stage shut down early, the other four burning longer to make up the difference.

But more then two days out, on April 13 one of two oxygen tanks in the unpressurised service module blew up, damaging the second tank which bled its life-giving contents to the vacuum of space. Immediately, the calm voice of Jim Lovell, understating the drama: 'Houston, we have a problem.'

Because the two tanks provided life-giving oxygen for the command module as well as gas for the fuel cells, the crew would quickly have no oxygen to breathe or electrical power to operate their spacecraft, call-sign Odyssey. They were two-thirds of the way to the moon and running out of ways to stay alive. The only option was to go in to the Lunar Module Aquarius and power it up, using it as a lifeboat all the way around the moon and back to earth.

'It's getting awful cold in here...'

About five hours after the explosion they fired the Lunar Module's descent engine to put them on a proper course for a slingshot back to earth – until that time they had been on a trajectory positioned for moon orbit,

Milestones

1970

APRIL 11
APOLLO 12 IS LAUNCHED FOR A POTENTIAL MOON LANDING AT FRA MAURO.

APRIL 13:
DISASTER STRIKES AS APOLLO IS SHAKEN BY AN EXPLOSION. 1970, APRIL 17: SPLASHDOWN FOR APOLLO 13.

not a free ride round and back courtesy of lunar gravity. After swinging round the moon – finding time to take pictures of the surface – they fired the descent engine a second time to speed up their return by nine hours. Built to support two men for less than two days, Aquarius now had to support three men for almost four days.

As they shut down everything to save power, temperatures dropped to freezing and ice formed inside the spacecraft. Shivering with cold their performance began to flag, lack of sleep began to sap their attention. When the carbon dioxide of their exhaled breath threatened to poison them the crew worked with Mission Control to jury-rig a makeshift adapter using filters from the command module in Aquarius.

LEFT • Left: The blast that shook Apollo 13 blew off the side panels of the service module, as seen when the command module separated shortly before re-entry.

'Goodbye Aquarius – and we thank you.'

For more than three days the crew limped home – numb with cold, almost no electrical power and in a spacecraft never before flown in this condition. After getting back in the command module and sealing the hatch with Aquarius they cut loose the damaged service module, photographing a hole in its side caused when the tank exploded. Then, with a final 'Goodbye Aquarius – and we thank you', they separated and came down to a splashdown only four miles from the USS Iwo Jima.

A triumph of determination to survive and to recover from a near disaster, Apollo 13 will always be the mission remembered not only for its crew but for the stoic group of flight controllers in Mission Control who kept it going. Caustic flight director Gene Kranz was supposed to have rallied the troops with the famous phrase 'Failure is not an option'. He never did say those words – but he might just as well have done.

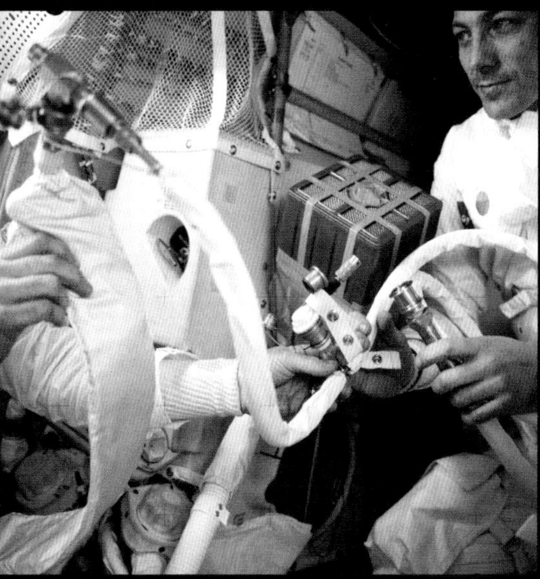

LEFT • Left: Jack Swigert connects air filtration pipes to a makeshift filter box strapped up with duck tape.

LEFT BELOW: • TV pictures stream in to Mission Control before the explosion that sapped oxygen supplies. Left and bottom: Flight controllers and astronauts huddle over consoles monitoring events aboard the spacecraft.

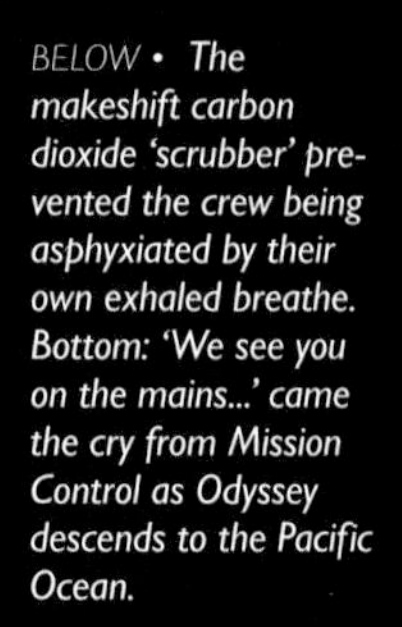

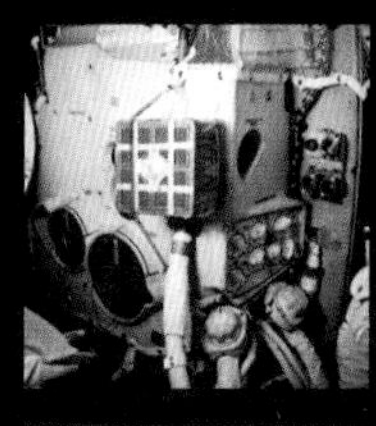

BELOW • The makeshift carbon dioxide 'scrubber' prevented the crew being asphyxiated by their own exhaled breathe. Bottom: 'We see you on the mains...' came the cry from Mission Control as Odyssey descends to the Pacific Ocean.

HEADING FOR THE HILLS

Swansong for a veteran as Alan Shepard puts the first wheel tracks on the moon.

Milestones

1971

JANUARY 31: LAUNCH OF APOLLO 14 FOR THE FRA MAURO HILLS, THE ORIGINAL TARGET FOR APOLLO 13.

FEBRUARY 5: LUNAR MODULE ANTARES LANDS AND REMAINS FOR 33.5 HOURS.

1971 FEBRUARY 9: SPLASHDOWN FOR APOLLO 14 WITH 94.5LB OF LUNAR ROCKS.

After the near disaster to Apollo 13, several changes were made to the spacecraft after an enquiry board discovered the oxygen tank that exploded had been damaged in ground tests. It also heard how the wrong specification for electrical control switches had exposed the wiring to higher voltages than it had been designed for. But just in case it should happen again, an extra oxygen tank was installed – on the opposite side of the service module.

Not for more than nine months did Apollo fly again, with Mercury astronaut Alan Shepard – the first American in space – commanding Apollo 14 and rookie astronauts Stuart Roosa and Ed Mitchell. It was great triumph for Shepard. Planned for 1963, his Mercury 10 orbital flight had been cancelled and he was given

BELOW • Shepard about to sink a core probe into the surface.
RIGHT • Strong sunlight makes it difficult to navigate a route across surface features.

command of the first Gemini flight, but in 1964 Shepard was diagnosed with Meniere's disease, an ear infection that grounded him until 1968. Now he was back with a vengeance.

Apollo 14 got off on January 31, 1971, but the mission was plagued with a series of minor problems. After departing earth orbit for the moon their was a problem getting the docking equipment to latch the Apollo command module to the Lunar Module and once in moon orbit the Lunar Module's computer became erratic, threatening to abort the landing attempt.

Rickshaw rolling

The principle task during the first space walk was to set up a set of scientific instruments that would continue to send back data about the moon for many years. Shepard also used a special 'thumper' device to set off a set of small charges on the lunar surface, the vibrations being picked up by a string of geophones – an array of small microphones strung out across the surface to measure the seismic waves and provide data about the structure of the soil. Shepard and Roosa also left a small mortar on the surface that would be remotely controlled from earth after they left the moon, firing small explosive charges and sending shock waves through the outer layers registered by a seismometer.

Because the crews of Apollo 11 and Apollo 12 had difficulty carrying all the tools they needed for getting moon samples and chipping off pieces of rock to bring back for analysis, Apollo 14 had a special handcart, dubbed a 'rickshaw', leaving the first wheel tracks in the lunar dust. Shepard and Mitchell set off on a long hike during the second EVA that should have taken them up the flanks of a crater called Cone. They had difficulty in the bright sun finding their reference and were unaware that they were in fact quite close to their objective.

'Miles and miles and miles!'

With their serious work done, Shepard and Mitchell had several minutes having fun. A serious golfer, Shepard took out a special six iron and two golf balls, sending at least one of them 'miles and miles and miles', in his estimation. In reality the balls only went about 300 yards! Then Mitchell took a lunar scoop handle and instead of leaving it on the surface as intended, proceeded to 'start the first lunar Olympics' by propelling it through the air like a javelin.

Unknown to anyone outside NASA, Dr Mitchell had been active in the scientific research of psychic connections and during quiet periods of rest in the Lunar Module he conducted experiments with two Universities in the US on mind-reading tests. Concentrating with test subjects on earth, the results were far higher than could have been expected through sheer chance...

ABOVE • *Photographs of the surface taken by the orbiting Apollo provided useful information for later missions.*

LEFT • *Most of the first of two space walks was taken up with laying out a set of experiments.*

LEFT • *On the airless moon shadows are deep black and sunlight is glaringly bright.*

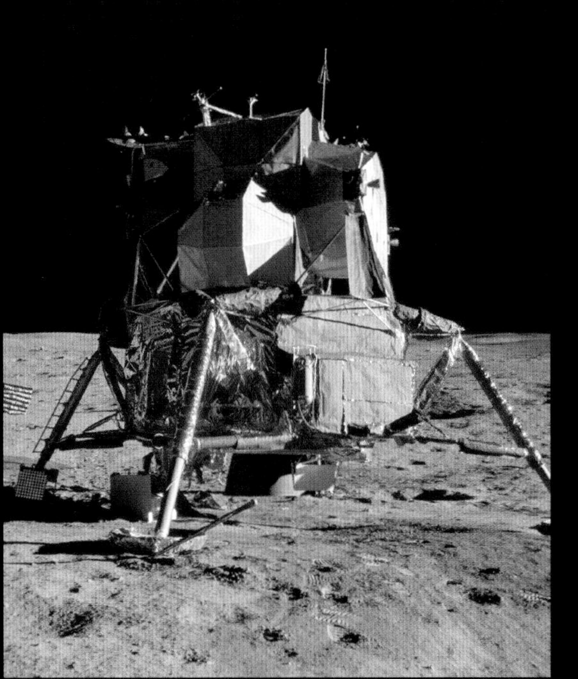

THE ELECTRIC

Big science begins

The first four moon landing missions had been planned to stay little more than a day and a half on the lunar surface, with no more than two periods working outside gathering samples. All that changed with

MAIN IMAGE: The two moon walkers drive their buggy to the slopes of Hadley Rille, a collapsed lava tube formed when great lakes of basalt oozed across the lunar surface nearly four billion years ago.

INSET: (left to right) Scott, Worden and Irwin.

Milestones

1971

JULY 26
ASTRONAUTS SCOTT, WORDEN AND IRWIN ARE LAUNCHED ON APOLLO 15 CARRYING THE FIRST MOON ROVING VEHICLE.

JULY 31
DAVE SCOTT BECOMES THE FIRST PERSON TO DRIVE A CAR ON THE MOON.

AUGUST 2
APOLLO 15 LUNAR MODULE FALCON LEAVES THE MOON WITH 169LB OF SAMPLES.

AUGUST 5
APOLLO COMMAND MODULE ENDEAVOUR SPLASHES DOWN IN THE PACIFIC OCEAN.

MOON BUGGY

a series of upgrades and improvements made to the spacecraft and the equipment the astronauts could use on the moon. These were in every respect, full geologic survey missions of selected sites on the surface and the last three Apollo missions to the moon were to reap undreamed of treasures that would keep the world community of scientists busy for decades.

Vital to it all was the Lunar Roving Vehicle, or LRV, to the astronauts simply the 'rover'. Built by Boeing under the direction of NASA's Marshall Space Flight Center in

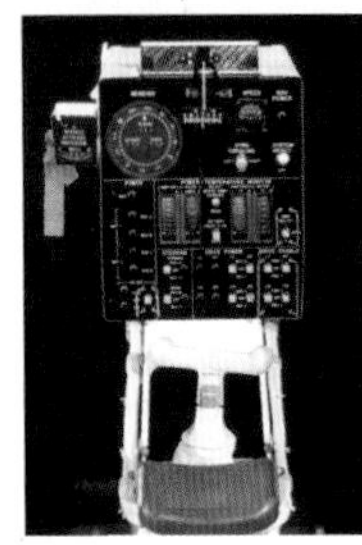

ABOVE: *An inertial navigation system on the moon buggy provided directional information.*

RIGHT: *Wire mesh wheels woven on the replica of a loom from Ancient Egypt fold over the chassis for stowage in the Lunar module*

BOTTOM: *Engineers practice unfolding the Lunar Roving Vehicle.*

BELOW: *When Apollo came home one of the parachutes collapsed increasing the speed of impact at splashdown.*

Huntsville, Alabama, the LRV would carry moon explorers several miles across the surface, exploring sites impossible to reach on foot. Powered by batteries recharged with solar cells, it carried all the cameras, geologic tools, sample bags and instruments they would need. Moreover, the LRV had navigation equipment taking them to pre-designated sites of interest and a TV camera to take the audience with them.

A hammer and a feather

Dave Scott is a natural born explorer and astronaut and a brilliant evangelist for moon science. He was a logical choice to command Apollo 15, the first of three J-series science missions, as they were known. In July 1971 Dave Scott and Jim Irwin piloted their Lunar Module called Falcon to a volcanic plain in the lee of mountains more than 7,000ft high. Swooping down across their peaks, Scott guided Falcon to a precision landing, prelude to three days of exploration. Across the valley floor they roamed, sampling rocks, photographing features and driving to the rim of a deep ravine that ran like a sinuous dried-up river bed miles across the lunar surface.

Galileo predicted that in a vacuum all objects would fall toward the ground at the same speed so to prove it, Dave Scott took a falcon feather and a geologic hammer in each hand and dropped them at the same time in full view of millions of people on earth. As predicted by the Italian astronomer 400 years ago, they fell at exactly the same speed and reached the surface together.

This was moon exploration as the science fiction writers had envisaged it. When Scott and Irwin dusted off, having driven 17.3 miles on three excursions, they had collected a record 169lb of samples, drilled down into the lunar surface and left temperatures probes to measure the amount of heat coming up from the centre of the moon. In addition they left the most comprehensive set of instruments to date and had geologically documented an area stretching several miles in each direction.

Falcon rested on the lunar surface for just over 66 hours and Scott and Irwin worked three shifts on the lunar surface gathering 169lb of rocks and samples. When they departed to rejoin Al Worden in Apollo, the liftoff was televised live from the rover, parked at a safe distance. But instead of the clipped voices coming from the moon at this supreme moment of tension, just as the engine lit up to lift them free a tape recorder on board Falcon played the stirring Air Force march 'Off we go into the Wild Blue Yonder!' to both smiles and looks of incredulity in Mission Control. This was one cool crew!

LAST MEN ON THE MOON

Road to the Highlands

In 1971 Apollo 15 had explored the Hadley-Apennine mountain range flanking a giant basin known as the Mare Imbrium and a sinuous gully caused by a collapsed lava tube. Apollo 16 would head for the rough highland areas far to the south because they were some of the oldest parts of the moon, a region known as Descartes. Like its predecessor, it would provide three periods of EVA and a roving vehicle to drive far from the landing spot.

Launched in April 1972, Apollo 16 carried John

BELOW: *Leaping clear off the surface, Dave Scott displays the liberation of only one-sixth gravity.*

"THIS WAS MOON EXPLORATION AS THE SCIENCE FICTION WRITERS HAD ENVISAGED IT"

ABOVE: *Apollo 16 blasts off for the Descartes region of the moon.*

RIGHT: *(left to right) Mattingly, Young and Duke.*

BELOW RIGHT: *Dust churned up by the roving vehicle throws rooster tails across the lunar surface.*

FAR RIGHT: *The crew of Apollo 17 pose for a publicity shot.*

BOTTOM RIGHT: *A fully laden moon buggy ready to roll to sites miles away from the Lunar Module.*

Milestones

1972

APRIL 16
ASTRONAUTS YOUNG, MATTINGLY AND DUKE ARE LAUNCHED ON APOLLO 16.

APRIL 27
APOLLO 16 RETURNS WITH 208LB OF MOON SAMPLES

DECEMBER 7
APOLLO 17 IS LAUNCHED AT NIGHT WITH CERNAN, EVANS AND SCHMITT.

DECEMBER 19
APOLLO COMMAND MODULE AMERICA SPLASHES DOWN IN THE PACIFIC OCEAN.

Young, Ken Mattingly and Charlie Duke to an accurate touchdown in a sequence that now seemed routine. But it was far from that and every mission had its fair share of glitches. This was real space flight - dangerous and unpredictable. The gremlins struck again shortly after Lunar Module Intrepid separated from the mother ship Orion in lunar orbit. A computer problem held things up for six hours but Intrepid finally made it to the surface at the start of a three-day campout in one of the most rock-strewn sites visited by Apollo.

A big rooster tail

Toward the end of the first EVA John Young took the rover on a wild ride across the surface – just to see how it would perform across the rough, undulating surface pock-marked with craters and small rocks. Charlie Duke kept up the dialogue worthy of a Grand Prix commentator: 'Man, you're really bouncing around'. And from Houston: 'Is he on the ground at all?' To which, replied the not altogether reassuring voice of Duke: 'He's got two wheels on the ground – a big rooster tail (of dust) out of all four wheels and as he turns he skids. The end breaks loose just like on snow. Come on back John. Man! I'll tell you. Indie's never seen a driver like this!'

Back to the valleys

On December 7, 1972, Apollo 17 lifted off at night in a spectacular blaze of light and sound, visible all the way up the eastern seaboard of the United States as it climbed into orbit. At the launch site was special guest and former black slave Charlie Smith, alive during the American Civil War of 1861-65. Aboard the mother-ship astronauts Cernan, Evans and Schmitt were heading for the Taurus-Littrow valley on the eastern flank of a giant basin known as Mare Serenitatis.

Before launch a sign had been hung on a work platform close to the Lunar Module, call sign Challenger, 'This May be the Last But it will be Our Best'. And it was. Accompanying Cernan down to the valley, Jack Schmitt was the only trained geologist to visit the moon, a place reserved for him when Joe Engle was asked to vacate that assignment. The Lunar Module was on the surface for 75 hours and Cernan and Schmitt drove the rover a total distance of 22.4 miles on three expeditions lasting a total of more than 22 hours.

Benediction

At the end of the third EVA, Cernan paused to reflect on Apollo, asserting that 'As we leave the moon at Taurus-Littrow, we leave as we came and God willing as we

ABOVE: 'Gene' Cernan rests after a moon walk, his suit coated with moon soil sticking like graphite dust.

LEFT: Cernan (left) and Harrison Schmitt, the only professional geologist to reach the moon.

BELOW LEFT: Lunar tracks from the moon buggy that will endure for centuries on the airless moon.

BELOW: A camera on the Lunar Roving Vehicle captures the moment of the last liftoff from the moon.

shall return, with peace and hope for all mankind.' And from geologist Jack Schmitt several weeks later back on earth: 'The record of Apollo, I believe, is a record of the audacity of Man to understand the moon. The record of his use of that understanding is just beginning...From this larger home we move to greet the future.'

In Houston during the flight of Apollo 17 a group of school children from many nations around the world had been on exchange visits. A small piece of moon rock was given to the representative of each country to take back with them to their own places of learning a token of the future envisaged for humanity when mankind journeys to other worlds together.

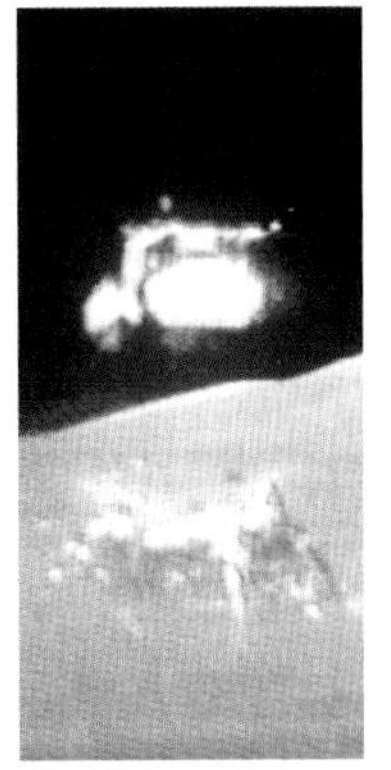

THE MOON UNVEILED

A global effort

Scientists learnt a lot from the exploration of the moon. None of which could have been discovered without exploring its surface and returning to earth samples for analysis in laboratories around the world. Apollo involved scientists and engineers from several countries and lunar samples are available to scientists everywhere. Each year they gather together to discuss their findings and piece together the jigsaw of what the moon is made of, how it was formed and why it is there.

World in collision

Earth's moon is larger in comparison to the size of its parent planet than any other, being one-quarter the diameter of the earth. Because of this, scientists say the earth and the moon is a bi-planetary system, each rotating around their common centre of mass. It has been known for several centuries that on average a cubic mile of the earth is around twice as heavy as a cubic mile of the moon. Now we know why and the answer lies back in the origin of the planets and the solar system.

Around 4.6 billion years ago the sun's disc of cooling debris came together in clumps that formed the planets. Lots of debris was left over, with some mini-planets in erratic and highly elliptical orbits and some that smashed into each other, breaking bits off that later came together to form other bodies. So it was that shortly after the earth formed, and while it was still cooling down, a massive object struck the earth and chunks of it flew off into space, re-forming over millions of years. Trapped by earth's gravity, it became our moon – half consisting of pieces of earth's light outer crust and half from the original impacting object.

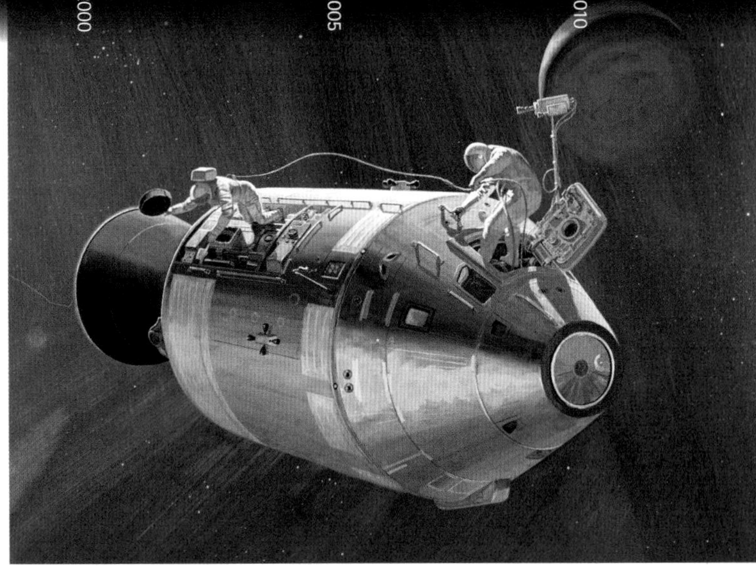

THIS PAGE CLOCKWISE FROM ABOVE: *The Apollo spacecraft on Apollo's 15-17 carried a battery of science instruments for scanning the lunar surface. On the way back to earth an astronaut retrieved film cassettes. Scanning the surface with radio waves mapped structure and chemistry while lunar samples added vital information to understanding the moon's origin.*

OPPOSITE TOP: *The moon formed after a giant impact struck the cooling earth more than 4.6 billion years ago. Debris formed into our moon through gravitational attraction.*

OPPOSITE BOTTOM: *Only one face is visible to observers on earth.*

LEFT: *Locations of lunar landing sites by mission number.*

MIDDLE LEFT: *Instruments left by Apollo astronauts helped map the moon's interior.*

FAR LEFT: *Detailed scanning from lunar orbit maps the height of surface features.*

A majestic dance

As if locked to the earth, the moon turns once on its axis every 28 days, exactly the time it takes to make one orbit of our planet. Hence, the same face always points toward us. Until the space age nobody knew what the far side looked like and it was found to be very different to the near side, with very few of the giant lakes of dried lava that make up the dark blotches seen on the nearside from earth.

Until Apollo nobody knew the age of the moon and in which order its features were laid down but now the moon is known to be almost as old as the earth and, when it was formed after that giant impact, it was very much closer to the earth.

Because of that the earth originally spun a lot faster a 'day' lasting only a few hours. Because the moon is slowly drifting away at the rate of 38mm each year, eventually the earth's spin will have slowed down, one 'day' lasting 27 hours long. These detailed measurements are made continuously by shining laser beams at reflectors left at each Apollo landing site.

The bottom line...

By bringing back rock samples, lunar explorers provided scientists around the world with material 99% of which is older than the oldest rocks on earth. This is because the moon died out quite early in its history while the earth has continually evolved geologically, churning and turning over most rocks older than 3.9 billion years. With a different isotope to earth rocks, moon samples unlock the very early history of the solar system hidden from geologists on earth. Now scientists have a view of the moon very different to the one we all see in the sky.

Who owns the moon?

Spacecraft launched by Russia, the United States and several other space-faring countries have landed or crashed into the moon and most have carried the flags of their host countries. But no one country has the right to sovereignty over earth's nearest celestial body because the 1967 Outer Space Treaty specifically defines the moon as the 'province of all mankind'. It also contains a protocol that prohibits any one nation from unilaterally exploiting its resources, but no-one has signed up to that!

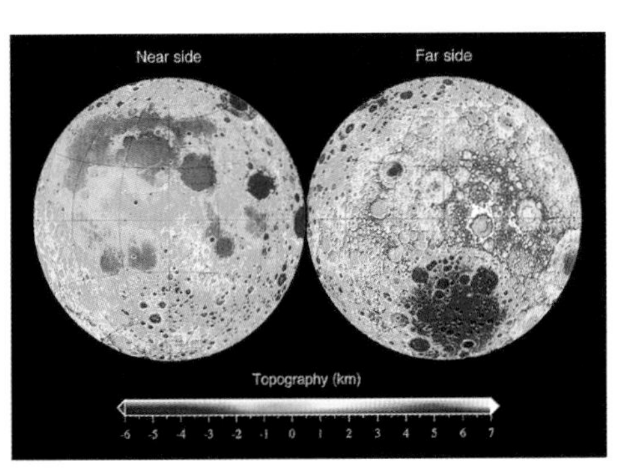

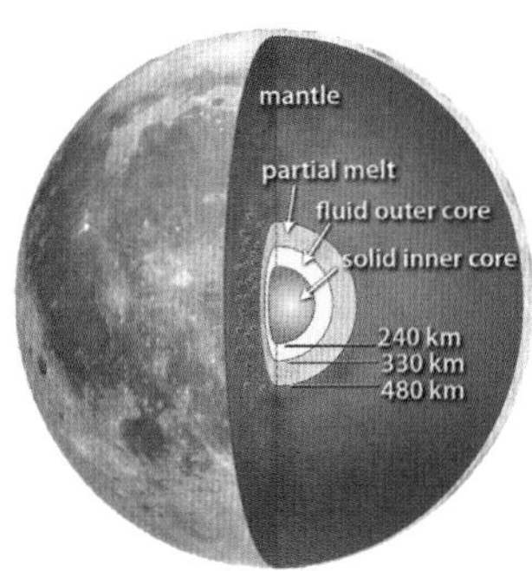

Conspiracy Theories

Some people say astronauts never went to the moon. What of the evidence?

For many people, the sheer scale of Apollo was unrealistic and from the early 1970s a range of conspiracy theories emerged to challenge the claim that NASA actually landed on the moon. Some theories were without substance or evidence of any kind, such as claims that it was simply too big a deal to pull off, that the flag placed on the airless moon was stiff and appeared to be blowing in the wind, and that the astronauts could never have survived the radiation of outer space. Other theories were backed by what their proponents believed was evidence of a conspiracy, such as fake photographs and wrong sunlight angles.

The scale of a deception

More than 400,000 people in several hundred companies worked for 12 years to support the Apollo programme and its moon missions. The vast majority of workers that would be essential to any conspiracy was outside the control of the government and had little or no reason to lie. Moreover, a million people saw with their eyes the launch of Apollo 11 and several hundred thousand more people watched the other missions leave the Kennedy Space Center for space. Where did those rockets go if not to the moon?

A damning indictment of conspiracy theorists is the Russians themselves. With a highly complex tracking and communication system, they watched and tracked Apollo missions leave earth and journey out to the moon and back, picking up telemetry and radio communication all the way out and back. The race to the moon was undertaken in the middle of the Cold War and the Russians would have dearly loved to have pinned a conspiracy on the Americans. They, and other astronomers with powerful telescopes, photographed

TOP CENTRE: ***Light balance levels confirm the correct sun angles.***

TOP LEFT AND ABOVE • ***Rock samples carry unique 'signatures' quite different from earth rocks.***

CENTRE AND BELOW: ***Instruments left at the last five landing sites continued to send data to earth for eight years. Laser reflectors are still used by scientists around the world. The Lunar Roving Vehicle provided TV coverage of the crew lifting off the surface.***

Apollo heading away from earth while amateur radio enthusiasts tuned in to voice links.

At tracking stations in Spain and Australia foreign nationals working with Apollo engineers were the primary communications link through giant dish antennas picking up signals as the earth rotated and the United States was, for a period each day, out of sight of the moon. Many other countries were involved and all were able to track the missions for themselves. Information did not come from NASA alone.

Why conspire to deceive?

Some conspiracy theorists believe the United States was so shaken by early Soviet space successes close to earth that President Kennedy challenged the Russians to a moon race simply because it was so far away and nobody really believed the Soviets could get there to disprove any US claim. Others believe that the number of accidents and potential catastrophes that plagued early space operations made a successful string of moon landings impossible to achieve in the time.

Evidence from the moon

Every day scientists fire laser beams at the moon, at places NASA says its astronauts laid out retro-ranging reflectors, tray-size arrays of light reflectors that bounce the laser light back to earth. Measuring not only the precise distance between the earth and moon they also track the motion of the continental plates on earth. Powerful evidence that they were deliberately placed there.

For eight year after the moon landings, arrays of scientific instruments sent data to earth on radio frequencies collected by antennas around the world, information about the solar particles striking the surface, about moonquakes deep below the surface and about trace gases drifting through fissures and cracks. Powerful evidence that they were laid out exactly as the astronauts claim.

And finally, there is evidence from the moon rocks themselves. Several thousand scientists in several hundred research laboratories around the world borrow moon rocks to conduct tests and carry out research. The rocks themselves are unlike those found on earth and each has the unique isotope, a 'fingerprint' pointing to their origin on a very different world in a completely different place.

ROOMS WITH A VIEW

NASA's first laboratory in space

Some say NASA stands for 'Never A Straight Answer'! It certainly did when they were annotating this prelaunch poster showing the Skylab in cutaway. No prizes for spotting the spelling mistake.

Milestones

1969

JUNE 19:
US AIR FORCE CANCELS ITS OWN SPACE STATION CALLED MOL AND TRANSFERS EXPERIMENTS TO SKYLAB.

JULY 22:
NASA ADDS AN APOLLO TELESCOPE MOUNT TO SKYLAB FOR STUDIES OF THE SUN.

AUGUST 8:
MCDONNELL DOUGLAS RECEIVES A CONTRACT TO CONVERT TWO S-IVB STAGES INTO A SKYLAB WORKSHOP CONFIGURATION.

1972

JANUARY 18:
ASTRONAUTS CONRAD, KERWIN AND WEITZ SELECTED FOR THE FIRST SKYLAB MISSION

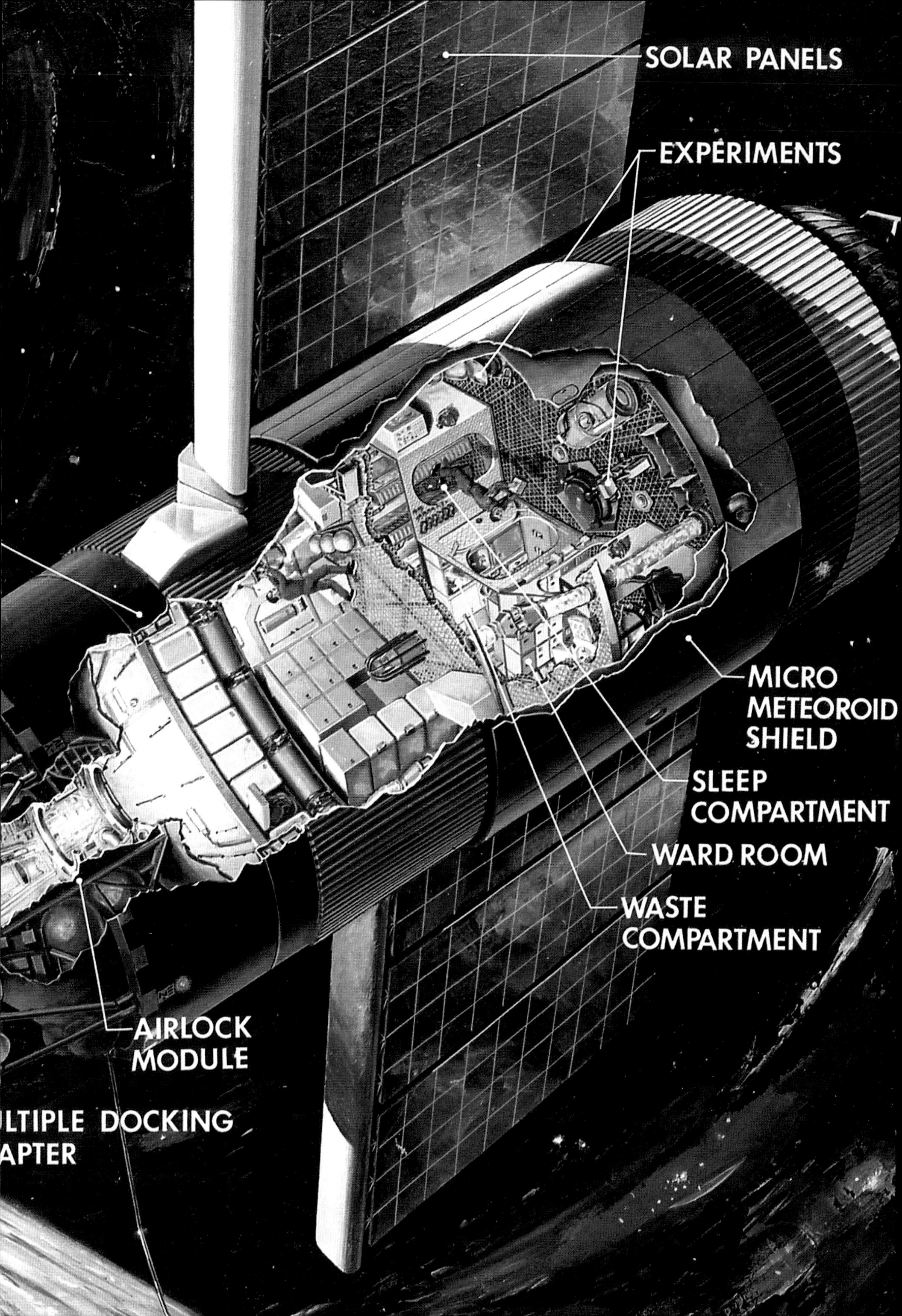

SOLAR PANELS
EXPERIMENTS
MICRO METEOROID SHIELD
SLEEP COMPARTMENT
WARD ROOM
WASTE COMPARTMENT
AIRLOCK MODULE
ULTIPLE DOCKING
APTER

Long before the first landing on the moon in July 1969, NASA had been planning to use Apollo spacecraft and Saturn rockets to build scientific research bases on the moon and orbiting space stations around earth. The moon missions ended in 1972 without plans for surface bases being approved but NASA was allowed to fly its Skylab space station as an intermediate step between Apollo and the Shuttle. It was to be a place in earth orbit where science and technology could benefit from research in a weightless environment.

Experiments conducted in Gemini during the mid-1960s indicated great value could be had from the weightless world of orbital flight. In a science known as 'materials processing', semiconductor crystals could be grown five times bigger than anything on earth. The booming microelectronics industry of the 1960s and '70s would benefit greatly from mini-factories in space. Also, new and much improved alloys could be made in weightlessness, offering promise of stronger but lighter metals for use on earth. Finally, there was a lot to be learned about the human body from its reaction to weightlessness and physicians were keen to study astronauts as guinea-pigs for health cures on earth.

Living and working in space

Space stations had been a part of NASA's long range plans since the agency was formed in October 1958. The grand plan for moon landings had shifted priorities to the race with Russia, but now the space programme could resume the path planned for it from the beginning. This time though, all the big rockets and the reliable space vehicles necessary for routine access to space were there – built up through the Apollo programme. It was a great opportunity to build on past success and take techniques and experiments pioneered by the two-man Gemini spacecraft and apply them to space stations in orbit.

Research in an orbiting space station would call for the crew to remain in space for many weeks at a time, perhaps several months. It would need a large volume accommodating all the compartments necessary for equipment to support several astronauts at a time. Places to work and conduct experiments, a washroom for personal hygiene, a place to eat – preferably together – and a space for sleep and relaxation. The design of a space station would be very different from a craft like Apollo, designed to spend relatively short periods in space and impose in the short term privations impossible to sustain for several weeks.

Using the leftovers

In planning its first space station NASA looked at equipment and facilities readily available and decided to use the empty third stage of a Saturn V rocket, fitted out with all the essential compartments and work stations. The three-stage Saturn V had been designed to blast 45 tons of spacecraft to the moon. The third stage was necessary to nudge itself and its payload into earth orbit and to fire again to send Apollo to the moon. But if the weight above the first two stages could be drastically reduced, there would be no need for the rocket motor on the third stage and the empty tank would provide an empty volume to serve as a space station. The crew would be launched separately on a smaller Saturn IB of the type used to send Apollo 7 into space in October 1968, and more crews could follow, getting a lot of useful work out of the adapted rocket stage – named Skylab.

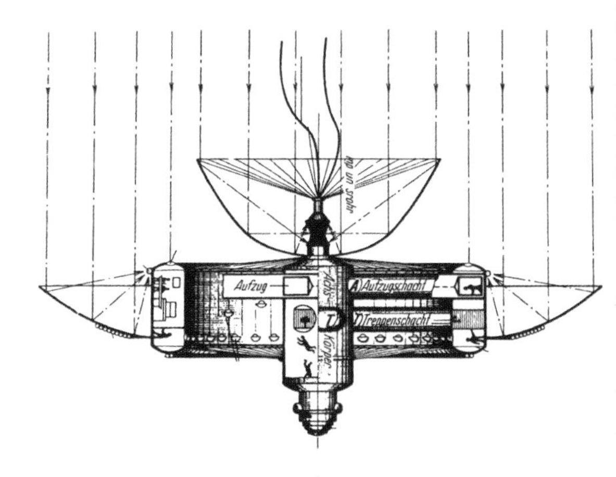

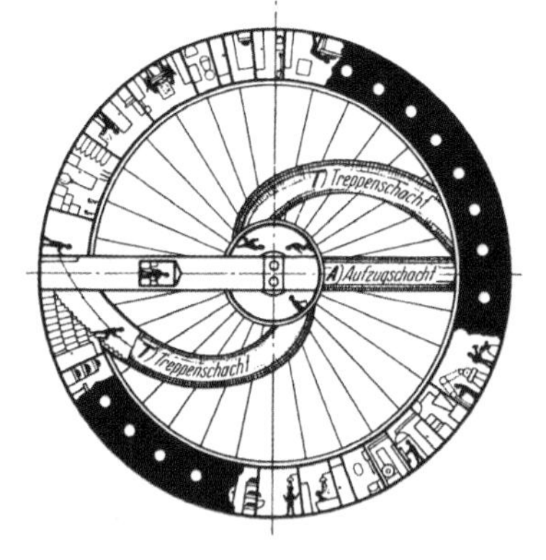

RIGHT • Hermann Noordung designed this space station in 1929, inspiring a generation of rocketeers.

BELOW • Naval vessels played a large part in maintining contact with Skyab. Far right: The Saturn S-IVB stage adapted into the Skylab space station

The last Saturn V carries NASA's Skylab space station into orbit, May 14, 1973.

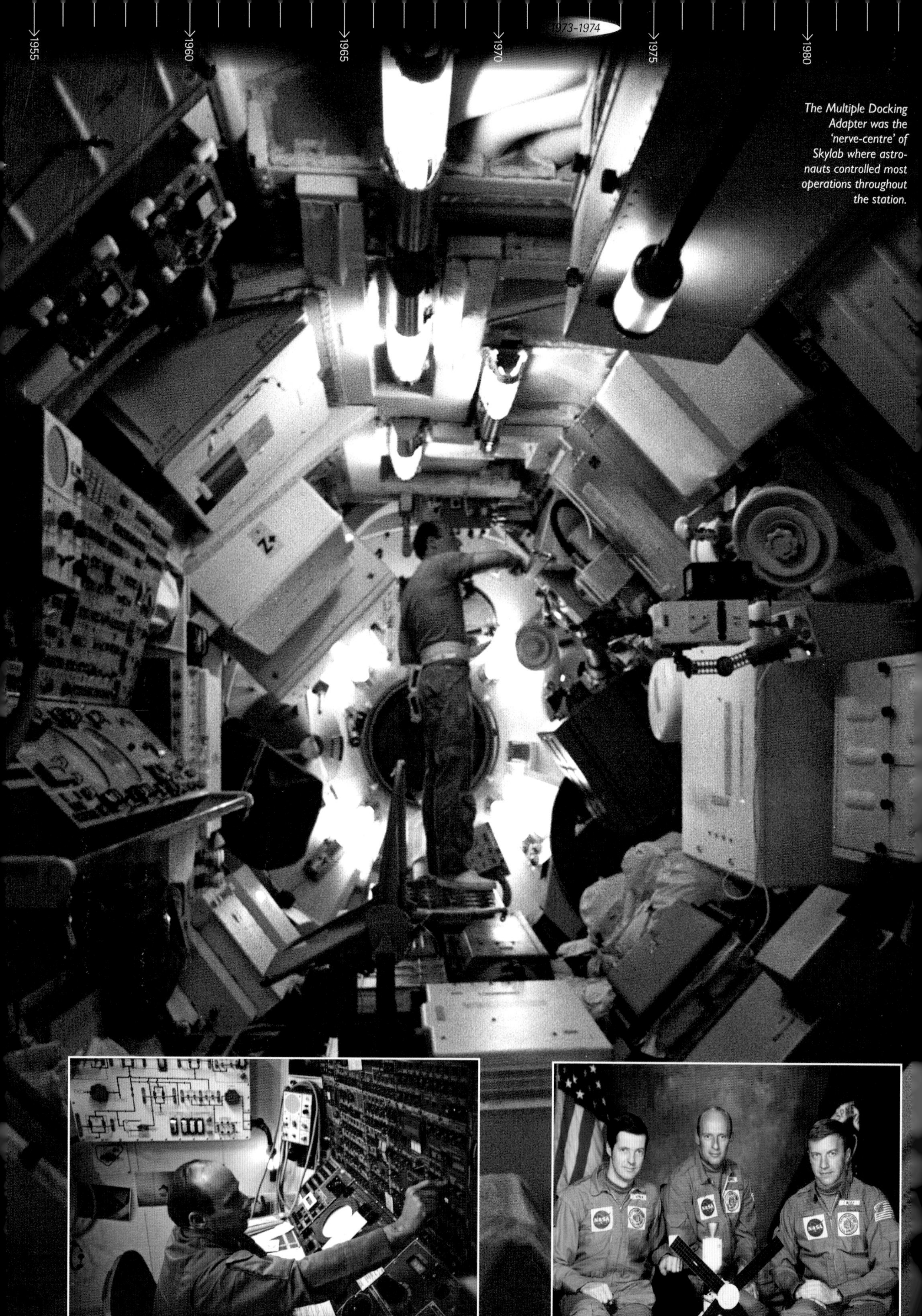

The Multiple Docking Adapter was the 'nerve-centre' of Skylab where astronauts controlled most operations throughout the station.

Checking solar telescope operatiog procedures on a ground simulator.

(left to right) Weitz, Conrad and Kerwin performed a rescue job after they arrived at the crippled Skylab.

An Orbital Workshop

Because Skylab was built up from the shell of a Saturn V third stage it had a lot of room inside, the interior volume comprising 10,000 cubic feet, about the amount of space in a three-bedroom house. It was separated into two sections by an open mesh divider that became a ceiling for the living area below and a floor for the work area above. A cutout section in the centre allowed the crew to move freely between the two sections. The stage itself had an interior height of 46 feet and a diameter of 22 feet, with a weight of just over 30 tons.

The astronauts would breathe an earth-like mixture of nitrogen and oxygen although Skylab's internal pressure was little more than one-third that at sea-level on earth. This is about the pressure of the atmosphere at 27,000ft but the astronauts could breathe normally because they had the same amount of oxygen as on the ground. It is possible to significantly reduce the amount of nitrogen a person breathes without damaging the body's respiratory system, but a sea-level oxygen level is essential for health and normal operation of the brain. However, some level of nitrogen is necessary for flights lasting longer than two or three weeks.

SKYLAB ORBITAL WORKSHOP

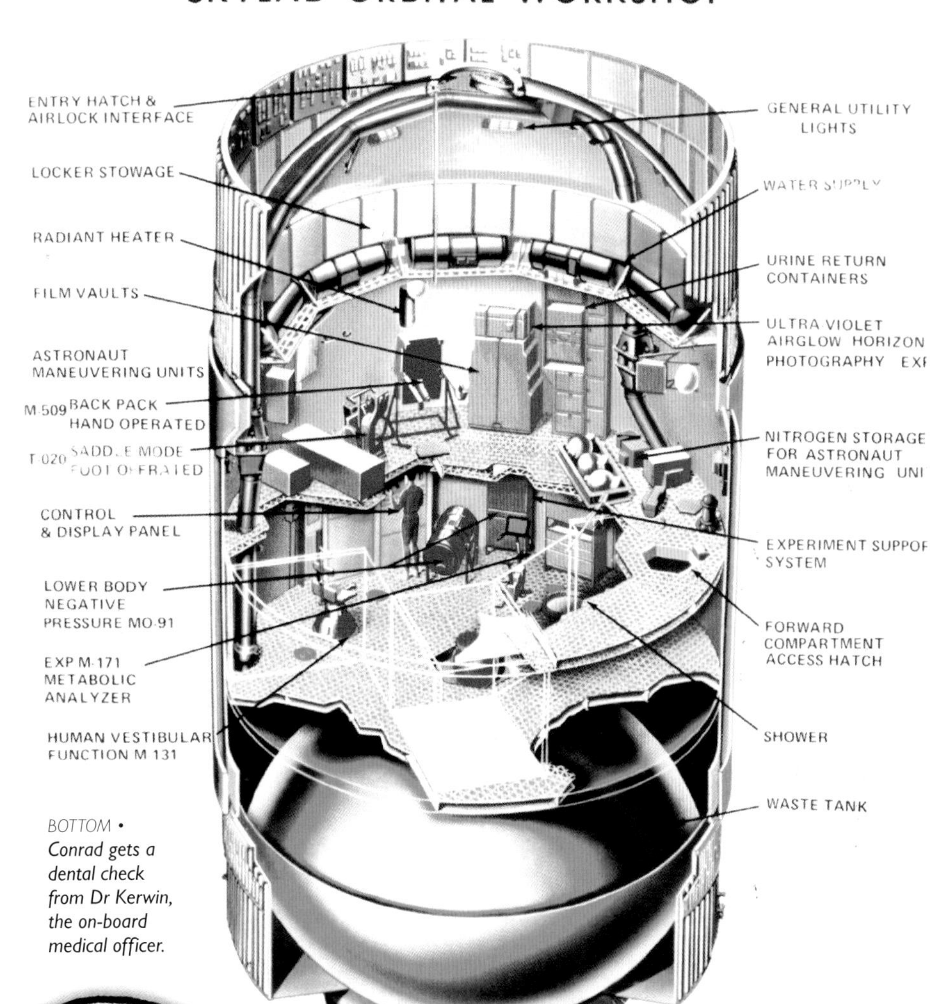

Building it bigger

The forward section of Skylab supported an Airlock Module, a Multiple Docking Adapter, and what NASA called an Apollo Telescope Mount. The Airlock Module was 17 feet long and 10 feet in diameter at the widest point, to which was attached the Docking Adapter, extending the length by a further 17 feet. Connected together, they provided an additional 1,700 cubic feet of space – about the size of a small living room.

The Apollo telescope Mount originated when NASA wanted to adapt a Lunar Module to carry very large solar arrays and a battery of telescopes for studying the Sun. Eventually, it became a separate structure in its own right, a box shape almost 15 feet long and 11 feet across with four long solar panels for electrical power. It was not pressurized but astronauts would use the Airlock Module to make a space walk and retrieve film cassettes from the telescopes. For launch all the forward modules would be protected by a shroud, with the Telescope Mount pivoted forward over the Docking Adapter until the shroud was jettisoned. After reaching orbit the Telescope Mount would rotate to one side and deploy its solar panels.

BOTTOM • Conrad gets a dental check from Dr Kerwin, the on-board medical officer.

The grand plan

Skylab had originally been scheduled for launch in 1972, between several Apollo moon flights planned for 1973 and 1974. But Apollo's 18, 19 and 20 were cancelled and the earlier flights spread out, so Skylab was put back to follow the last moon flight and launch in early 1973. The first two stages of the Saturn V that would have launched Apollo 20 were assigned to place the Skylab space station into orbit, followed a day later by the first crew launched on a Saturn IB.

Milestones

1972,

FEBRUARY 28: REVIEW OF ALL SKYLAB SYSTEMS SUPPORTS A PLANNED LAUNCH DATE OF APRIL 30, 1973.

APRIL 3: SKYLAB GROUND SUPPORT EQUIPMENT BEGINS TO ARRIVE AT THE KENNEDY SPACE CENTER.

JULY 26: SKYLAB MEDICAL EXPERIMENTS ALTITUDE TEST CONDUCTED BY ASTRONAUTS.

AUGUST 22: STAGES FOR SATURN IB TO LAUNCH FIRST SKYLAB CREW BEGIN TO ARRIVE AT KENNEDY SPACE CENTER.

DECEMBER 15: LAST TWO SKYLAB SOLAR ARRAY WINGS ARRIVE AT KENNEDY SPACE CENTER.

Milestones

1973

MAY 14:
SL-1, THE SKYLAB SPACE STATION, IS LAUNCHED FROM THE KENNEDY SPACE CENTER BUT IMMEDIATELY RUNS INTO TROUBLE.

MAY 25:
SL-2 IS LAUNCHED CARRYING ASTRONAUTS CONRAD, KERWIN AND WEITZ.

MAY 26:
ASTRONAUTS ENTER SKYLAB AND DEPLOY AN UMBRELLA-LIKE HEAT PROTECTION SHADE.

JUNE 7:
CONRAD AND KERWIN CONDUCT A SPACE WALK TO FREE THE JAMMED SOLAR ARRAY.

JUNE 19:
CONRAD AND KERWIN PERFORM A SECOND SPACE WALK TO RETRIEVE FILM CASSETTES.

JUNE 22:
SL-2 CREW RETURNS HOME AFTER A MISSION LASTING 28 DAYS 49 MIN 49 SEC.

Lift-off – into trouble!

The 98 ton Skylab space station was launched from NASA's Kennedy Space Center on May 14, 1973, into clear blue skies but little more than a minute into the flight it was already in trouble. About 63 seconds after liftoff one of the two solar arrays was torn free from the side of the Orbital Workshop, carrying with it a micrometeoroid shield that doubled as thermal protection from the full heat of the Sun's rays. Soon after reaching orbit temperatures inside the Workshop began to soar, reaching 125 deg F.

Moreover, debris from the torn shield fouled the second solar array which was prevented from fully opening. Virtually uninhabitable due to high temperatures and with almost no electrical power from the sole remaining array, plans to send up the first crew were cancelled. Teams at the Manned Spacecraft Center in Houston, and at the Marshall Space Flight Center in Alabama, began developing procedures whereby the first crew could attempt to deploy both the snagged solar array and a thermal shield.

'Pete, Paul and Joe Space Repairs'

Over the next several days, a plan developed that was bold and audacious. The first crew would chase after Skylab and before docking to it, an astronaut would stand up in the open hatch of the Apollo command module and using a 10 ft pole try to pry loose the debris holding back the damaged solar array. After docking the crew would enter Skylab and deploy through a small airlock designed to carry experiments, an umbrella-like shade that once pushed through to the outside could be unfurled to act as a sunshade over the bare hull of the exposed Workshop.

Distributing mock business cards declaring themselves available as the 'Pete, Paul and Joe Space Repair' team, Conrad, Kerwin and Weitz arrived at the Kennedy Space Center for a launch that took place on May 25, 10 days late but on track to pull off one of NASA's biggest in-space repair jobs of all time. Arriving at Skylab less than eight hours after launch, Weitz found it impossible to cut free the debris restraining the solar wing and then there were difficulties docking but they finally made it almost 15 hours after launch.

Fit for purpose

After a much needed sleep the crew entered the Orbital Workshop and deployed the umbrella-like device in hot and uncomfortable conditions. Almost immediately the temperatures began to fall but not until the fourth day were they down to 90 deg F. At risk had been the food, prepackaged and refrigerated for three periods of occupation. The first crew was scheduled to remain aboard for 28 days – doubling the longest NASA manned flight to date – followed by two missions each expected to last 56 days.

About half way through their month-long stay aboard Skylab, the crew performed a space walk and this time managed to cut free the snagged solar array wing. Until this point, virtually all the electrical power had come from the four big solar arrays on the Apollo Telescope Mount, but now they could power up

experiments that previously had been put on hold. And, they could enjoy a few more comforts that needed electrical power! Then a few days before returning to earth, Conrad and Weitz did another space walk to retrieve film from the solar telescopes.

They returned to earth after 28 days in space, the longest manned space flight to date, their repair job done, useful scientific work completed and experiments set up for the next crew in line. Above all, they had saved a potentially disastrous catastrophe that but for the presence of humans would have rendered the space station lifeless.

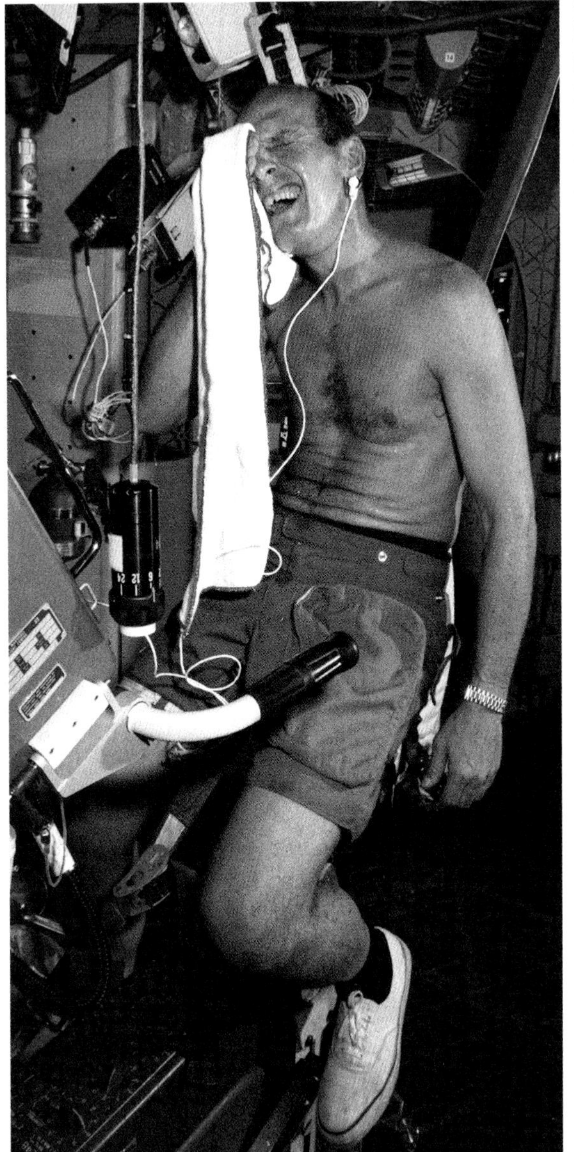

OPPOSITE PAGE AND TOP RIGHT • *With one solar array wing torn away, and carrying a special blanket-like solar sunshade, Skylab was nursed back into operation for a full mission and more science than planned.*

LEFT, CENTRE AND BELOW • *Pete Conrad on the exercise cycle to compensate for muscle loss in weightlessness and carrying out science tasks.*

FAR LEFT • *Vance Brand and Don Lind would have flown a Skylab rescue mission to put it back in use, had the Shuttle been available in 1979 as planned.*

BELOW AND FAR LEFT • *Spacewalks were the only way to retrieve film cassettes from the solar telescopes.*

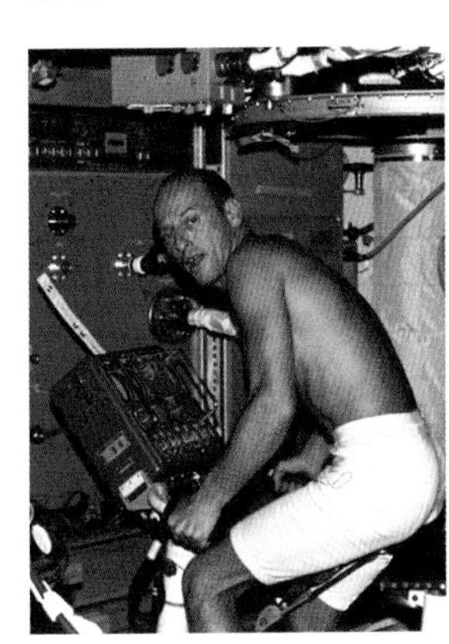

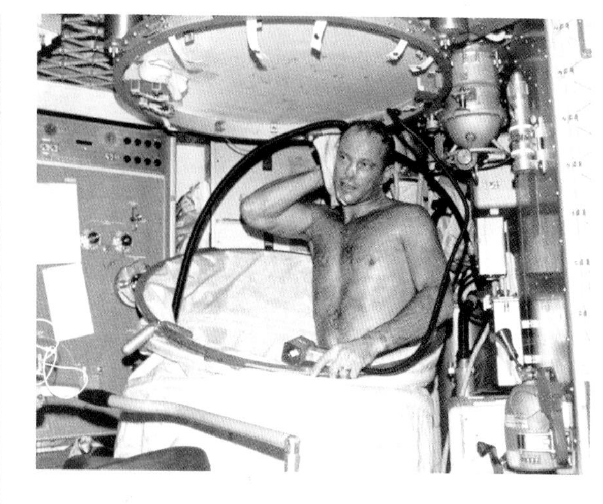

ABOVE • *Studies into the disorientating effects of weightlessness being conducted by Pete Conrad.* TOP RIGHT• *Jack Lousma gets a shower, with water taken away by a device similar to a waterproof vacuum cleaner.* ABOVE RIGHT • *Carr, Gibson and Pogue, the last crew aboard the final Skylab flight.*

Milestones

1973,

JULY 28: SL-3 CARRYING BEAN, GARRIOTT AND LOUSMA IS LAUNCHED TO SKYLAB.

AUGUST 6: SKYLAB SPACE WALK WHERE GARRIOTT AND LOUSMA ERECT A PERMANENT SUNSHADE.

AUGUST 24: SECOND EVA FOR GARRIOTT AND LOUSMA TO REPLACE GYROSCOPES.

SEPTEMBER 22: THIRD EVA WHEN BEAN AND GARRIOTT RETRIEVE FILM CASSETTES.

SEPTEMBER 25: SL-3 RETURNS AFTER 59 DAYS 11 HRS 9 MIN 4 SEC.

NOVEMBER 16: SL-4 LAUNCH WITH ASTRONAUTS CARR, GIBSON AND POGUE.

NOVEMBER 22: FIRST EVA WITH GIBSON AND POGUE TO REPAIR EQUIPMENT.

DECEMBER 25: SECOND EVA WITH CARR AND POGUE TO CHANGE FILM IN SOLAR TELESCOPE.

DECEMBER 29: THIRD EVA TO PHOTOGRAPH COMET KOHOUTEK.

1974

FEBRUARY 3: FOURTH EVA TO RETRIEVE FILM CASSETTES. 1974, FEBRUARY 8: CREW RETURN HOME AFTER 84 DAYS, 11 HR, 16MIN.

Rollout for a rescue

Skylab was unoccupied for 36 days, until the SL-3 crew comprising Alan Bean, Owen Garriott and Jack Lousma headed for orbit on July 28, 1973. They were to remain aboard Skylab for nearly two months, double the duration of the previous mission. High at launch, confidence was dashed when Apollo had trouble with its thrusters. Struggling to catch up with the orbiting laboratory, the crew managed to dock with Skylab just over eight hours into their flight.

Thrusters are a vital means of maintaining control of the spacecraft in orbit. While docked to Skylab it was powered down but once activated for return to earth, anything could happen. So NASA rolled out the next Apollo designated to fly the third mission in a hurry and prepared for an early launch to bring the crew back. But it was not needed. It would await its turn carrying the third crew on the final visit to Skylab.

The space bug looms

About half the astronauts that go into space get space sickness – disorientation causing nausea – which usually takes a few days to subside. All three crewmembers got it and a space walk to erect a permanent sunshade was delayed a few days. The umbrella-device put out by the crew of the previous mission was good for what it was but NASA wanted them to put out a bigger shade to do a better job. It did and the temperatures gradually fell back to normal levels. Life aboard Skylab was a routine unlike previous space flights. Their day was fixed around the Houston clock, awakened at 6.00am after 8 hours sleep, 2 hours 30 minutes for meals during the day and sleep beginning at 10.00pm. In between, time was divided among the various experiments into processing new materials, earth observation, medical tests and experiments, operations with the solar telescope mount and a variety of other activities.

The design of Skylab's interior layout and the colours selected by NASA were on the advice of Raymond Loewy, famous for having designed the Coca Cola bottle, the Shell and BP logos and Studebaker cars, among other things. NASA wanted as normal an environment as possible and it worked. After two months in space, the crew returned home with a highly successful mission accomplished.

Skylab swansong

The final mission to NASA's orbiting laboratory began on November 16, 1973 with the launch of SL-4 carrying Gerald Carr, Ed Gibson and William Pogue. At 84 days it was to be the longest of all, providing boundless opportunity for science and experimentation. When they got there they found three crewmembers already in residence. Three dummies dressed up by the previous crew The final flight provided a unique opportunity to photograph the Comet Kohoutek, making its first appearance after 150,000 years. With perfect timing, the Skylab crew was able to take advantage of their unique position high above the earth. When they returned, the nine Skylab astronauts had provided scientists with 175,000 photographs of the Sun and more than 46,000 pictures of the earth, conducted a total of almost 42 hours on 10 space walks and supported three missions totaling 171 days and 13 hours of space flight. Not bad for a programme made up from leftovers.

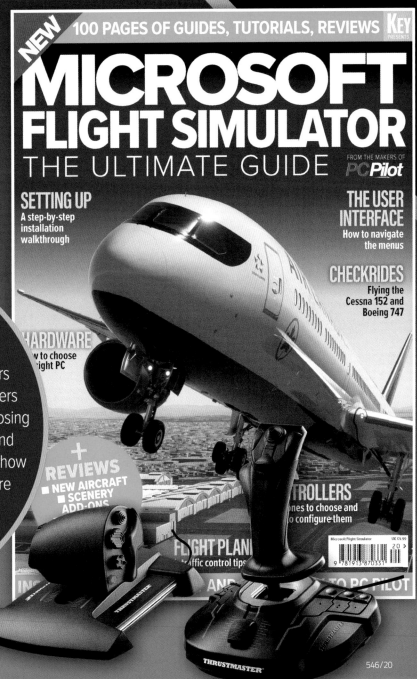

EAST MEETS WEST

Détente in space

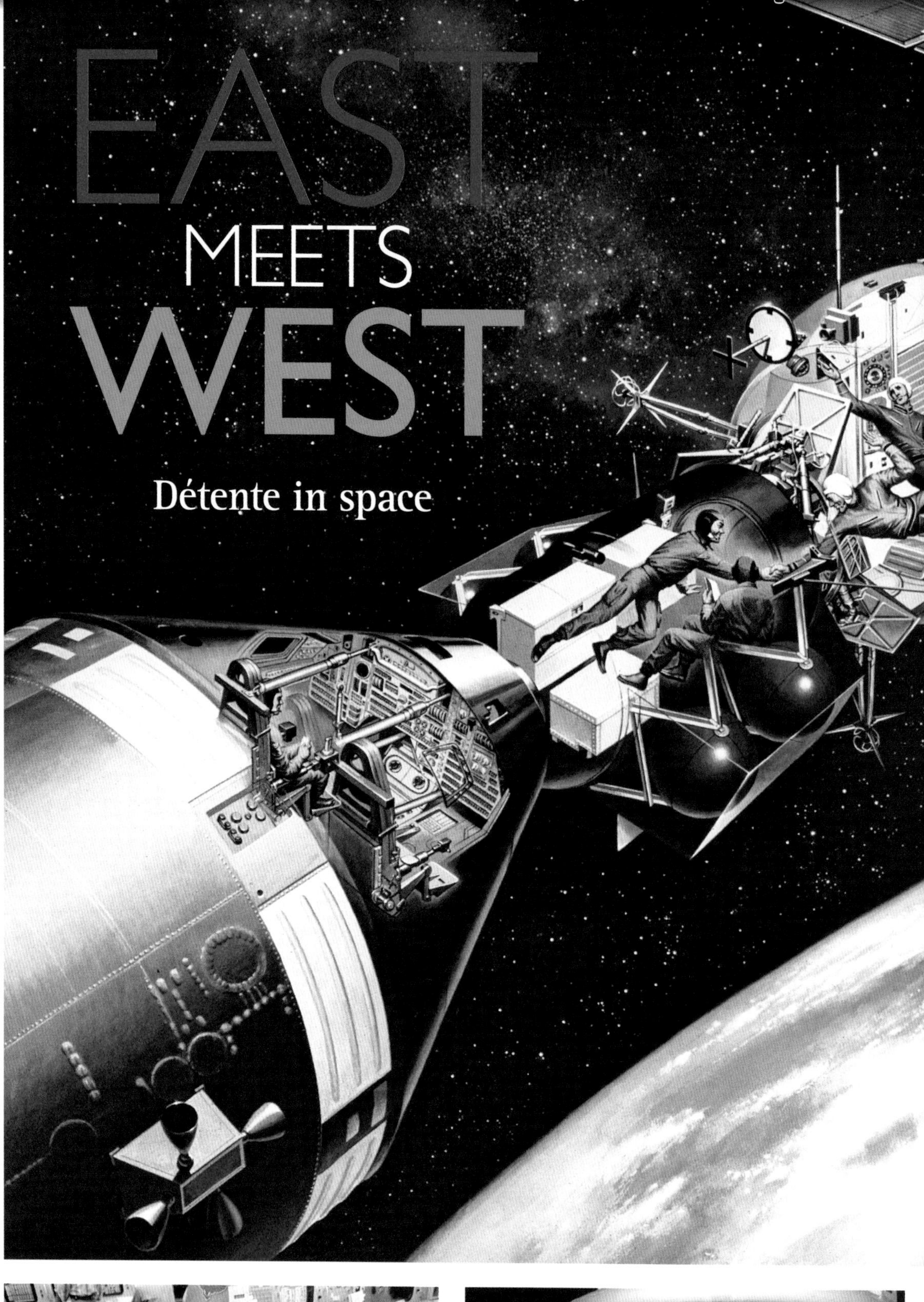

RIGHT: A cutaway of the docked Apollo spacecraft (left) and the Soyuz (right) with the docking module connecting the two.

Milestones

1963

JULY 23
PRESIDENT KENNEDY SOUNDS OUT THE RUSSIANS ON CO-OPERATION.

SEPTEMBER 18
KENNEDY BRIEFS NASA BOSS ON POSSIBLE MERGER OF APOLLO GOAL WITH RUSSIA.

NOVEMBER 22
KENNEDY IS ASSASSINATED AND ALL DISCUSSIONS OVER CO-OPERATION ARE OFF.

1971

JANUARY 16
A JOINT DOCKING FLIGHT WITH RUSSIA IS PROPOSED.

1972

MAY 24
AGREEMENT ON THE APOLLO SOYUZ TEST PROJECT.

ABOVE: The last Saturn IB is rolled out to the pad for the launch of Apollo.

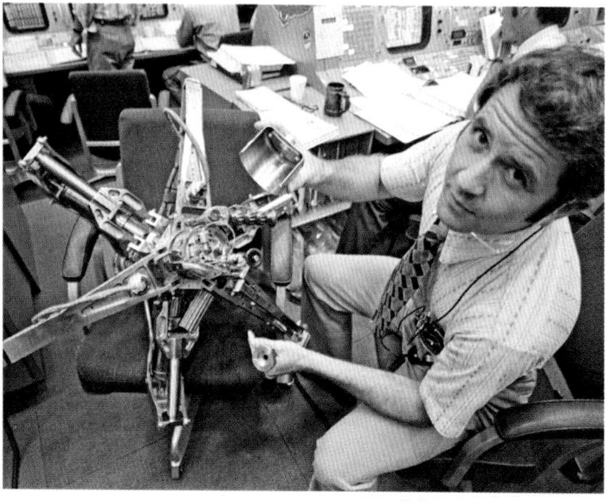

RIGHT: An engineer displays the components of the new docking probe to link together the two spacecraft.

FAR RIGHT: Testing the androgynous docking system specially developed for the joint mission.

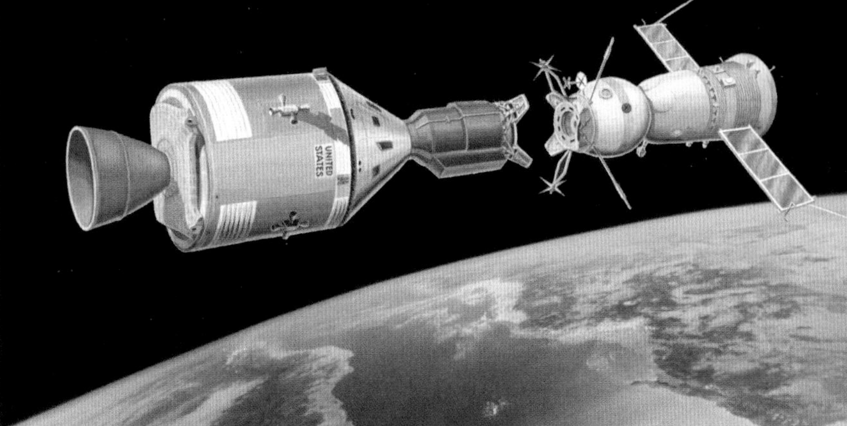

ABOVE: *The Apollo spacecraft nudges toward the Russian Soyuz in this artist's view.*

LEFT: *Crews and back-up personnel gather for a photo-shoot.*

BELOW *(left to right) Slayton, Stafford and Brand pose before their Saturn IB.*

Fast forward six years and with the moon race won, interest in some sort of joint space flight was renewed. President Nixon saw it as a way for political advantage, détente with the Russians. Slowly plans were made and feelers put out, for possible co-operation using the Apollo spacecraft and a Soyuz capsule to seal a new age of friendship. It fitted well with the arms control talks and the first international deal to limit the missile race.

Bridging the gap

In 1970 and 1971 astronauts and cosmonauts visited each others' countries in a friendly exchange. From this grew awareness that perhaps such a linkup was indeed practical. But there were technical difficulties. The Russian Soyuz had a nitrogen/oxygen atmosphere at sea-level pressure while Apollo had a pure oxygen environment at one-third atmospheric pressure.

Then there was the problem of connecting the two very different spacecraft. Each country had its own docking mechanisms and they were incompatible. There would need to be a Docking Module, much

ABOVE: *Stafford (left) and Slayton lead a group in front of St Basil's Cathedral alongside the Kremlin in Moscow*

BELOW: *An Apollo mock-up where Leonov (left), Stafford (right) and Slayton rehearse procedures.*

Back in 1963, President Kennedy began secret discussions with Soviet space chiefs over the possibility of abandoning NASA's Apollo moon programme in favour of a joint deal with the Russians. In a letter to NASA's deputy boss, Hugh Dryden, dated July 23 that year, he ordered the space agency to get ready for such a deal. Horrified, many senior managers at NASA objected to the concept at that time in the Cold War, but on September 18 Kennedy met with NASA boss James Webb and told him to keep the agency under control. He really wanted to make it happen. Two months later Kennedy had been assassinated and his successor, Lyndon Johnson, was having none of it. No more was said and the nation put its shoulder behind the moon race.

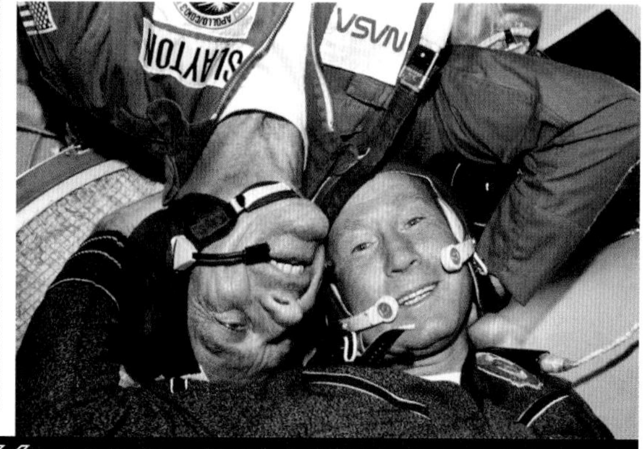

RIGHT (FROM TOP) : With no up or down in space, Leonov (right) and Slayton get in close for a historic picture; Stafford and Leonov greet each other; Kubasov in Soyuz; Stafford exiting the docking module; Brand in Apollo.

"WELCOME ABOARD SOYUZ' 'MAN I TELL YOU, THIS IS WORTH WAITING 16 YEARS FOR!"

BELOW: Preparation for launching the Soyuz spacecraft.

Milestones

1975

JULY 15
SOYUZ AND APOLLO ARE LAUNCHED FOR A RENDEZVOUS IN SPACE.

JULY 17
THE TWO SPACECRAFT DOCK AND ASTRONAUTS SHAKE HANDS WITH COSMONAUTS IN SPACE FOR THE FIRST TIME.

JULY 19
APOLLO AND SOYUZ SEPARATE FOR INDEPENDENT MISSIONS.

JULY 21
SOYUZ RETURNS HOME TO A LANDING IN KAZAKHSTAN.

JULY 24
APOLLO SPLASHDOWN IN THE PACIFIC OCEAN, THE LAST APOLLO MISSION.

like an airlock, carried by Apollo to connect the two spacecraft. A common docking system would have to be designed and fitted to the front of the Docking Module so that it could latch on to the Russian ship. And last but certainly not least, there was the little matter of communication. Astronauts and cosmonauts would have to speak each other's language and be able to do that with technical conversation and communication to and from space.

Getting it together

But there was more to it than agreeing over whether it was a good idea, feasible but difficult to accomplish. Both sides were protective of their technology and opponents of the idea, both in America and Russia saw it as an excuse for spying on each others' technology. Then there was the matter of control. With Apollo controlled from Houston ad Soyuz from a command centre outside Moscow, who would be in charge? Gradually, the ground rules were written for a shared responsibility and plans for the first international space link-up began to take shape.

Building bridges

By 1972 the go-ahead had been given and astronauts and cosmonauts went on exchange visits to each other's country. While teaching Russians how to cook burgers in Houston, Americans learnt from Russians how to drink Vodka in Moscow. And when US space officials let their hair down and set off fireworks against the Kremlin wall on the 4th of July, the Russian secret police, although alarmed at first, understood the explanation and joined in – after all, they understood about revolutions!

The Soviets had suffered a disaster in June 1971 when cosmonauts Dobrovolsky, Volkov and Patsayev were asphyxiated after their Soyuz 11 spacecraft depressurized in space. They needed a big fix, the world sharing the pioneering dangers of exploration on the last frontier. As for the Americans, this would be the swansong for Apollo, coming more than a year after the last Skylab flight.

Veterans day

To command the last Apollo, NASA chose Tom Stafford, a veteran of Gemini and Apollo moon flights. Donald K (Deke) Slayton and Vance Brand, would also be on board. A veteran of the space programme, Deke Slayton had been a member of the first group of Mercury astronauts in 1959. At 51, he was making his first space flight, an event delayed when in 1962 it was discovered he had an erratic heart rate. A lengthy medical restored him to flight status for the Apollo Soyuz Test Project – the official name for the historic linkup in space.

The Russians chose Alexei Leonov to command the Soyuz spacecraft, partnered with Valery Kubasov, a rookie. Another veteran from the past, Leonov had made the first space walk in history, during the flight of Voskhod 2 in March 1965, and he had been in line to command the first

FAR LEFT *Apollo and its attached docking module as seen from Soyuz.*

LEFT *The Soyuz spacecraft viewed from Apollo.*

BELOW MIDDLE: *A mock-up in the National Air & Space Museum, washington DC.*

BELOW: *From the Oval Office, President Ford places a historic call to the spacecraft.*

moon landing – which never took place. Now, these men would make history and begin a process that two decades later would lead to full international co-operation in building the world's largest space station.

Launch day

The flight plan for Apollo-Soyuz had the Russians launch first, from Baikonur on July 15, 1975, followed by the Apollo spacecraft from the Kennedy Space Center on the last Saturn IB, 7 hours 22 minutes later. With Soyuz in a fixed orbit 143 miles above the earth, Apollo played chase – it had the reserves of maneuvering fuel which the Russian spacecraft did not – and so began two days of playing catch-up. The two spacecraft were together just over 51 hours after launch and they were docked within the hour.

Handshakes in space

Quite by chance and a quirk of the flight path, the historic hand-clasp was supposed to take place over Bognor Regis while the docked spacecraft were passing over the south coast of England. But it was delayed and the event took place over Metz, in France. It was July 17.

Leonov and Stafford were the first to shake hands, the culmination of several years training in each other's spacecraft on the ground and in simulators. Then it was the turn of the others to embrace and for just over seven hours they exchanged medals, presents and mementoes of their mission. All too soon it was time to undock, and for Soyuz to re-dock with Apollo in a test of the new 'androgynous' docking mechanism.

Soyuz remained in space a further three days carrying out some scientific experiments and photographing the earth below. During re-entry the Apollo crew misunderstood instructions to control the release of the parachutes, causing the spacecraft to swing violently. As attitude thrusters fired incessantly to control the motion, fumes were sucked in to the cabin causing Brand to pass out until restored when Stafford held an oxygen mask over his face. Then the self-righting bags failed and the command module remained upside down until righted by divers. An ignominious way to end the Apollo missions!

End of an era

When Apollo splashed down in the Pacific Ocean near Hawaii, it brought to an end three generations of US manned space vehicles – Mercury, Gemini and Apollo – all of which had splashed down in water. This was the last of those ballistic capsule flights and from now on astronauts would fly back from space in the Shuttle and land on runways. But not for nearly six years would another American go into space.

ABOVE RIGHT: *Leonov, an amateur artist, becomes the subject in this all-crew depiction.*

ABOVE LEFT: *(left to right) Slayton, Stafford, Brand, Leonov and Kubasov.*

INSET LEFT: *Highly collectable stamps depict the joint flight.*

BELOW: *Toxic fumes from vented rocket fuel affected the crew at splashdown.*

WINGS INTO

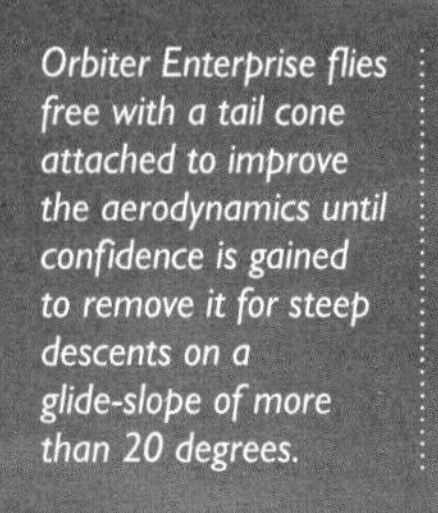

Orbiter Enterprise flies free with a tail cone attached to improve the aerodynamics until confidence is gained to remove it for steep descents on a glide-slope of more than 20 degrees.

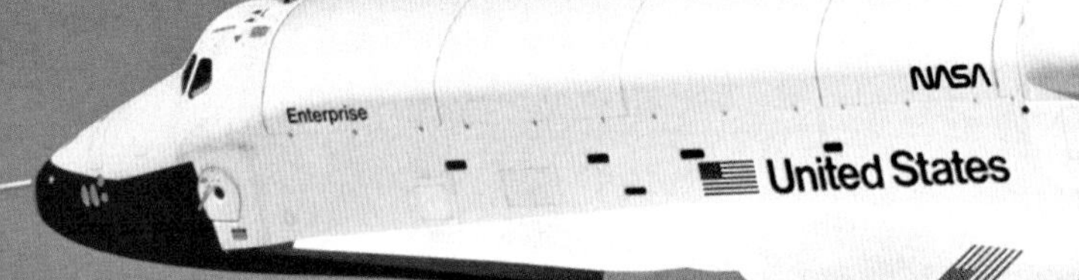

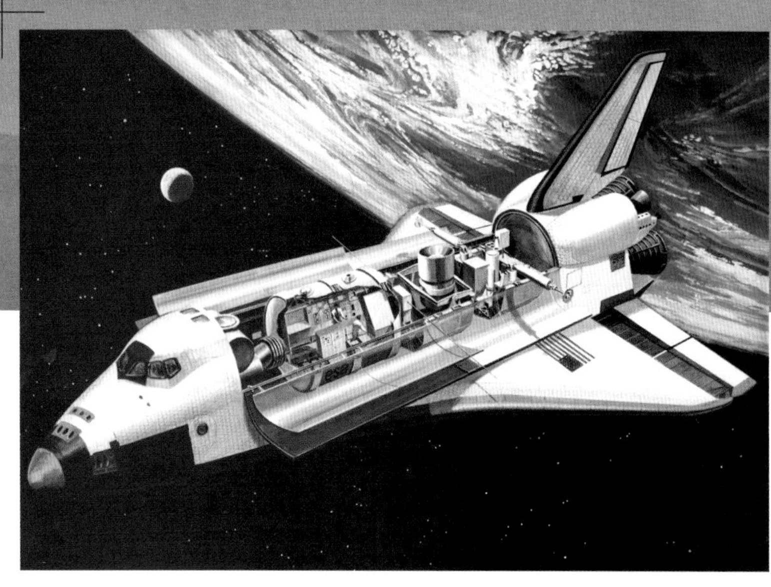

"IT WAS A LONG TIME COMING BUT THE SHUTTLE WAS AN UNPRECEDENTED CHALLENGE - THE FIRST REUSABLE SPACEPLANE"

ABOVE: The Shuttle was designed to carry scientific payloads, satellites for earth orbit and spacecraft destined for the planets, plus a European laboratory called Spacelab to continue the research begun by NASA's Skylab of 1973-74.

RIGHT: Above Star Trek enthusiasts were delighted when most of the cast from the TV series came to the rollout of Enterprise on September 17, 1976.

For more than 14 years NASA sent its astronauts into space in spacecraft that could be flown only once, on rockets thrown away during each launch. Now it was going back into orbit – with a spacecraft that had wings to fly back down through the atmosphere and land on a conventional runway.

A new Enterprise

They had wanted to call it Constitution and as a symbol of national pride the name could hardly have been better. But when Star Trek enthusiasts heard about that they lobbied NASA to change the name of its first Shuttle. When it was rolled out in 1976 it bore the name Enterprise, after the starship in the famous TV series.

NASA had been planning to replace its existing, expendable spacecraft with a reusable vehicle combining the assets of a launch vehicle and a spacecraft since the mid-1960s. Even before Neil Armstrong planted the first boot print in the lunar dust, engineering plans were appearing on drawing boards at NASA research centres and aircraft plants across the United States. Finally, in January 1972, President Nixon gave it the go-ahead and Rockwell got the contract to build the world's first reusable Shuttle.

SPACE

NASA's reusable Shuttle

In addition to launch pads at the Kennedy Space Center in Florida, the Shuttle was also to have been launched from a specially adapted pad at the Vandenberg Air Force Base in California. Enterprise was carried there to check the facilities' electrical and umbilical connections but the site was never used.

Milestones

1968
AUGUST 10
NASA'S ADMINISTRATOR FOR MANNED SPACE FLIGHT DELIVERS A SPEECH TO THE BRITISH INTERPLANETARY SOCIETY IN LONDON WHERE HE ANNOUNCES THE DECISION TO BUILD A REUSABLE SHUTTLE.

1969
JANUARY 31
NASA ISSUED CONTRACTS FOR INDUSTRY STUDIES ON A REUSABLE INTEGRAL LAUNCH AND RE-ENTRY VEHICLE, OR ILRV

APRIL 24:
NASA'S NEWLY FORMED SPACE SHUTTLE TASK GROUP DISCUSSED VARIOUS AEROSPACE VEHICLES.

1970
MAY 9
NASA SELECTED NORTH AMERICAN ROCKWELL AND MCDONNELL DOUGLAS FOR DEFINITION STUDIES ON A SHUTTLE DESIGN.

1972
JANUARY 5
PRESIDENT NIXON AUTHORIZED DEVELOPMENT OF THE SHUTTLE.

JULY 26
NORTH AMERICAN ROCKWELL SELECTED TO BUILD THE SHUTTLE.

1976
SEPTEMBER 17
SHUTTLE ENTERPRISE IS ROLLED OUT AT PALMDALE, CALIFORNIA.

1977
AUGUST 12
BETWEEN THIS DATE AND OCTOBER 26, NASA FLEW THE SHUTTLE ENTERPRISE OFF THE TOP OF A TOP OF A CONVERTED BOEING 747 TO EVALUATE ITS FLYING QUALITIES.

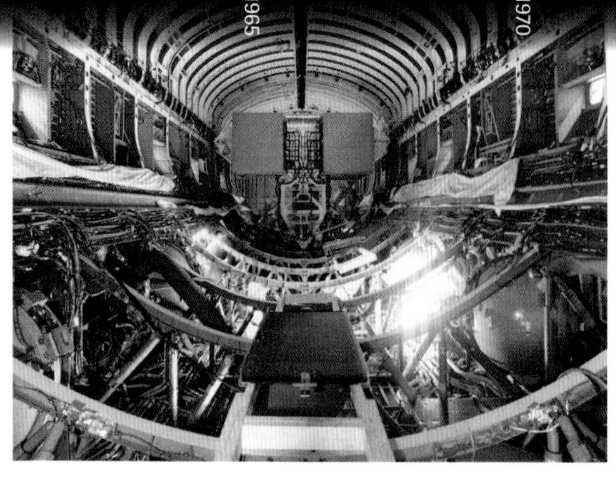

A new way to fly

The design of the Shuttle originated from research data extracted from the X-15 rocket flights in the 1960s and theoretical calculations. Northing like it had been made before – it was one of a kind – and there was nothing preceding it from which to gather experience. While the X-15 had provided valuable research data there was a big gap between the speed and temperatures it reached and those the Shuttle would experience. The X-15 had achieved speeds of more than 4,000mph, but the Shuttle was designed to re-enter the atmosphere at 17,500mph. The temperatures experienced on the X-15 were around 1,200 deg F, while those for the Shuttle would reach almost 3,000 deg F.

Instead of a single rocket engine like the X-15, the Shuttle Orbiter would carry three powerful rocket motors in its tail producing a total thrust of more than 1.1 million lb, supplied by liquid hydrogen and liquid oxygen propellants from a giant external tank to which it was attached. In turn, the external tank would be supported by two solid rocket boosters, each producing a thrust of almost 3 million lb. The boosters would burn out at about two minutes while the three Orbiter main engines would continue to burn almost to orbit. The tank would separate just short of orbital speed and fall back down into the atmosphere where it would burn up, leaving the Orbiter to push on into orbit using a 6,000lb thrust rocket motor each side of the tail.

THIS PAGE, CLOCKWISE FROM TOP LEFT:
The payload bay is capable of supporting more than 45,000lb lifted to space.; The pressurized crew compartment rests within the forward fuselage; Over the last 30 years advances have revolutionised the flight deck. The Orbiter's three main engines are installed in the Orbiter Processing Facility.

ABOVE: *In the Vehicle Assembly Building, an external tank is lowered between two stacked solid rocket boosters, after which the Orbiter will be lowered to the side of the tank.*

FAR LEFT: *The external tank is 27.5ft in diameter and 155ft long, covered with a spray-on foam insulation to protect the super-cold hydrogen and oxygen propellants.*

LEFT: *The external tank is fabricated from cylindrical structures of lightweight lithium-aluminium capable of holding the hydrogen as a liquid at -423 degree F, and the oxygen, in a separate vessel, at -297 degree F.*

Feeling the atmosphere

The first Shuttle Orbiter, Enterprise was not built for space flight. It had no heat protective tiles and it had no provision for the three main engines. It was used for air-launched tests off the top of a Boeing 747 out of NASA's Dryden Flight Research Center in California. The first three drop-tests were made with a cone fitted over the tail, both to smooth out turbulence on the tail of the 747 and to give it better handling qualities for these first flights down to the ground. The first drop-test took place on August 12, 1977, remarkably less than 20 years after Sputnik 1 and less than nine years after the last X-15 flight. While the Orbiter has been said to fly like a brick, it did well enough to justify removing the tail cone and flying the very steep descent all Orbiters would fly returning from space. The last two flights had the tail cone removed and dummy shapes installed replicating Orbiter rocket motor nozzles to realistically measure its handling qualities.

Getting ready for space

While Enterprise went off to be used for tests at various NASA facilities, the second Orbiter, named Columbia, was being built at Palmdale. It arrived at the Kennedy Space Center in March 1979 but it would be more than two years before it flew the first Shuttle mission. There was much work still to be done, testing its big main rocket engines, tweaking the design of heat protective tiles and preparing all the equipment necessary to launch. Many engineers doubted the Shuttle could achieve a safe flight into space and back, and some doubted it could ever live up to its design promises. The entire future of NASA's manned space flight programme was riding on the first flight, the first time a manned space vehicle had carried humans into space without first having been tested in orbit. The risk was colossal.

BELOW: *Columbia arrives at the Kennedy Space Center in March 1979 but will wait two years before it gets to fly.*

GENESIS OF A NEW SPACE AGE

After 20 years of flying capsules, now astronauts would ride on wings.

On April 12, 1981, the 20th anniversary of the flight of Yuri Gagarin, astronaut John Young and Robert Crippen were launched aboard Columbia from the Kennedy Space Center in Florida. It was the first flight in which a new generation of winged space vehicles would conduct missions in orbit – the first flight of NASA's long-awaited Shuttle.

Columbia shows the way

The dangerous job of the test pilot is frequently to climb into a new vehicle and take it into the air for the first time on a flight fraught with danger. Nothing like it had ever flown before and the Shuttle had never been tested fully assembled and with all engines and solid boosters firing. NASA estimated the possibility of a catastrophic loss as one in 100,000. Thirty years on, with two Shuttle tragedies from which to learn, that estimate of the danger on those early flights has been revised to a probability of one in nine. When asked today about that, Young is sanguine, saying he never believes in theoretical calculations anyway! For him it was, 'the greatest flying machine ever!'

The flight had its troubles but nothing to mar an otherwise successful mission. There were problems with sound waves at lift-off pounding on the Orbiter's wings and dislodging 16 tiles – but that could be solved for the next flight. And the Orbiter behaved just a little differently than engineers had predicted – but not so as to shake confidence in the overall design.

On orbit, Young and Crippen carried out a wide range of tests, checking all aspects of the Orbiter, and they conducted orbital manoeuvers with the rocket motors in the tail pods. Designated STS-1, the mission lasted just over two days and when they came back to earth it was to a perfect landing at Edwards Air Force Base in California, where there is plenty of runway for margins of error.

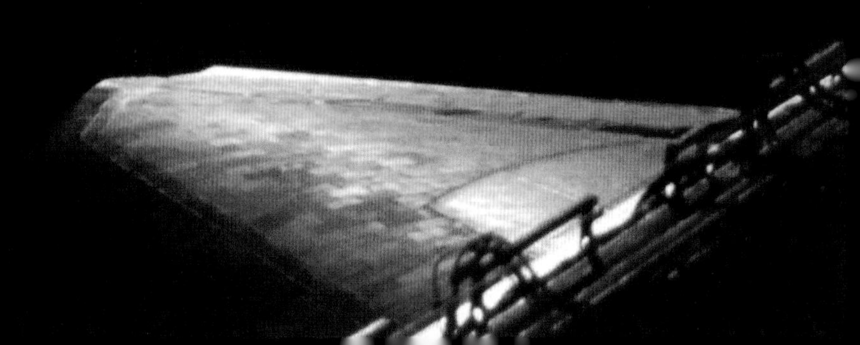

Sunset over the earth's limb, looking aft along the payload bay from windows on the flight deck.

“IT WAS THE FIRST FLIGHT IN WHICH A NEW GENERATION OF WINGED SPACE VEHICLES WOULD CONDUCT MISSIONS IN ORBIT - THE FIRST FLIGHT OF NASA’S LONG-AWAITED SHUTTLE.”

Veteran of Gemini and Apollo flights, commander of the Apollo 16 moon landing in 1972, John Young (left) and Robert Crippen pose for a publicity shot before their historic launch.

BELOW RIGHT • *Two years after arriving at the Kennedy Space Center, Columbia stands ready for its first flight in April 1981.*

Milestones

1981

APRIL 12: YOUNG AND CRIPPEN PILOT THE FIRST SHUTTLE, COLUMBIA, INTO ORBIT ON THE 20TH ANNIVERSARY OF THE FLIGHT OF YURI GAGARIN.

1982

NOVEMBER 11: THE FIRST COMMERCIAL PAYLOAD CARRIED BY THE SHUTTLE CAME ON THE FIFTH MISSION WHEN SATELLITES FOR CANADA AND A US BROADCASTER WERE RELEASED IN ORBIT.

1983

APRIL 4: THE SECOND FLIGHT RATED ORBITER, CHALLENGER MAKES ITS FIRST FLIGHT ON THE 6TH SHUTTLE MISSION WHICH ALSO SAW THE FIRST SHUTTLE-BASED EVA.

JUNE 18: NASA'S FIRST WOMAN ASTRONAUT, SALLY RIDE FLIES CHALLENGER ON THE 7TH SHUTTLE MISSION.

NOVEMBER 28: EUROPE'S SPACELAB FLIES ON COLUMBIA, NASA'S 9TH SHUTTLE FLIGHT.

1984

JANUARY 25: PRESIDENT REAGAN FORMALLY APPROVES DEVELOPMENT OF A SPACE STATION.

AUGUST 30: THE THIRD ORBITER DESTINED FOR SPACE MISSIONS, DISCOVERY MAKES ITS FIRST FLIGHT ON THE 12TH SHUTTLE FLIGHT.

1985

OCTOBER 3: FOURTH ORBITER, ATLANTIS MAKES ITS FIRST FLIGHT AS THE 21ST SHUTTLE MISSION.

The new race to space

It seemed that everyone wanted to get in on the act. Having shown the way, there were so many uses for the Shuttle, the only problem was it was taking longer to turn it around for the next mission. But that was OK, there were more Orbiters being built and soon they would be firing off the pad at the rate of a mission every two weeks. At least, that was the idea but it didn't quite work out that way. As flights progressed the difficulty of getting the Orbiter ready for its next flight never got any easier and a constant succession of minor technical problems delayed each mission. NASA pinned it hopes on more Orbiters joining the fleet and increasing the number of missions each year.

Next up was Challenger, converted to space flight from a series of structural ground tests it had originally been built for. Challenger made its first flight in 1983, joined in 1984 by Discovery and in 1985 by Atlantis. NASA wanted a fifth Orbiter but Congress refused the funds, voting instead to allow NASA to build up a set of major structural spare parts, just in case a flight Orbiter needed repair. Even with four Orbiters in the fleet, it was still proving difficult to fly more than eight or nine missions a year. But why the hurry to increase the number of flights? This was because a global industry was waiting for launch slots.

The Shuttle is moved more than three miles from the giant Vehicle Assembly Building, built in the 1960s for assembling Apollo-Saturn rockets. It is on a mobile launch platform carried by the worlds largest tracked land vehicle, the so-called crawler transporter.

Liftoff! Columbia heads for space, the culmination of a decade's work to build the world's first reusable spaceship.

“DESIGNATED STS-1, THE MISSION LASTED JUST OVER TWO DAYS AND WHEN THEY CAME BACK TO EARTH IT WAS TO A PERFECT LANDING AT EDWARDS AIR FORCE BASE IN CALIFORNIA.”

Left: *Just over 54 hours after launch, Columbia lands at Edwards Air Force Base, California.*

Transformation

Around the time the Shuttle started flying in the early 1980s, space technology had outgrown its original use as a tool for national prestige, as part of a space race driven by the Cold War. By the 1980s tensions between East and West were easing and there was less need for national competition and the political will to beat the Russians for the loyalty of uncommitted nations had bled away. A new Soviet leadership wanted to talk friendship and the space programme evolved into something very different from the objectives it had when President Kennedy announced the moon goal two decades earlier.

The new space race involved commercial companies wanting to launch communications satellites so that people in remote and Third World countries could gain access to telephone and telex facilities, achieving that from satellites at a fraction of the cost of wiring up dispersed and sparsely populated areas. Then satellite broadcasting pushed the use of satellites to transmit programmes to schools in Third World countries. This was also driven by the need to raise educational standards and train people how to improve crop yield, manage their natural resources and teach their children.

Commander of the second Shuttle flight, a veteran of X-15 flights, Joe Engle (left) had been replaced on the last Apollo moon mission to make way for a qualified geologist, Harrison Schmitt, but got to go into space, along with Richard Truly.

BELOW • The Florida peninsula is seen down to the Keys, with Cape Canaveral and the Kennedy Space Center half way up the eastern side. The Gulf of Mexico is to the left, the Atlantic Ocean to the right.

RIGHT • Looking down at the Nile Delta on Egypt's Mediterranean coast 200 miles below the Shuttle. Earth views play an important role in studying the planet and understanding its environment.

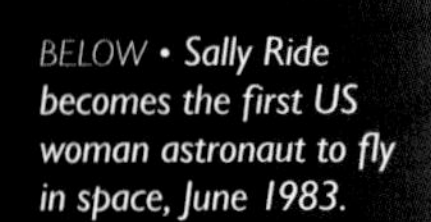

BELOW • Sally Ride becomes the first US woman astronaut to fly in space, June 1983.

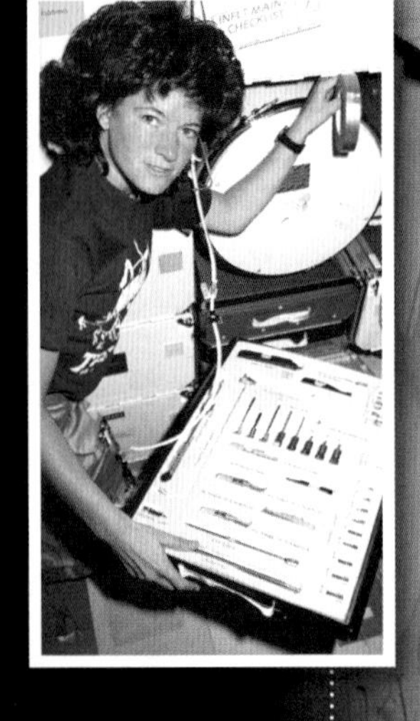

RIGHT • Several science experiments were conducted by Sally Ride during her space flight. Behind her is the EOS (Electrophoresis Operations in Space) machine, a co-operative venture with pharmaceutical company Johnson & Johnson to perfect a single-shot cure for diabetes.

WOMEN IN SPACE

It had been more than 22 years since NASA's Alan Shepard had been launched into space on a Redstone rocket as America's first astronaut, but Sally Ride made a big step for gender equality when she was launched into space aboard Challenger in June 1983. There had been hostile reluctance to allow women into a man's world and the test pilot culture of the early years forbade their consideration for such roles, a view typified by John Glenn who said at a hearing in 1963 as to whether women should be considered for space flight: 'men fight the wars and men fly the airplanes!'

It had taken a long time to get women selected but that chauvinism had been toppled in 1978 when Sally Ride was inducted to the astronaut corps, following Soviet cosmonauts Valentina Tereshkova, who flew in 1963, and Svetlana Savitskaya launched in 1982. Born in 1951, Dr Ride was among 8,000 people applying to be astronauts of whom 35 were accepted for training, six of them women. A physicist, she was also an accomplished tennis player but put her professional skills to work within the space programme, making a second flight aboard Challenger in 1984. Retiring from NASA in August 1987 she had served on the Rogers Commission looking into the loss of Shuttle Orbiter Challenger.

Other women selected in 1978, the first astronaut selection since 1969, included Anna Fisher (the first mother in space), Shannon Lucid, Judith Resnik, Rhea Seddon and Kathryn Sullivan. Since then many women have flown aboard the Shuttle and served as pilots too. In July 1984 Soviet cosmonaut Svetlana Savitskaya became the first women to walk in space, followed three months later by the first NASA female astronaut, Kathryn Sullivan, to do the same.

Of the almost 600 people who have been launched into space in the last 60 years, including the select few who soared there on the wings of the X-15, just over 60 have been women. This roll of honour includes Judith Resnik and Christa McCauliffe who died aboard Challenger during launch on January 28, 1986, and never made it all the way into orbit.

Buying a ride

Countries such as India, Indonesia, Mexico, Thailand, Brazil, Argentina and the remote regions of Canada, Australia and Arab states wanted satellites for this purpose. It was business in those places that pushed commercial space technology, rapidly adopted also by Americans and Europeans for increasing their living standards and recreational options through multi-channel TV and satellite telephone systems. Where it had been necessary to book a call to the United States in the late 1970s, by the late 1980s a caller in Boston, England, could pick up the telephone and dial Boston, Massachusetts.

Dwarfed by the sheer size of this Leasat communications satellite, astronauts from Discovery work on repairs after which it will be placed back in its operating orbit.

Commander Joe Engle configures the computers on the flight deck of Discovery during the 20th Shuttle flight in August 1985.

A jet-pack known as the manned maneuvering unit was flight tested by astronaut Bruce McCandless on a Challenger mission in February 1984.

Flying free and un-tethered from Challenger, Bruce McCandless (dubbed Bruce 'McCordless'!) moves away to a distance of 300ft from the Orbiter.

Erecting poles built up from a kit carried aboard Atlantis, astronauts practice how to assemble components of a space station.

BELOW • *In September 1984 a large solar array extending 105ft above Discovery was deployed to test panels designed for the space station.*

Buying a ride on the Shuttle with satellites for what became known as 'space applications' became a matter of national urgency – to raise the standards in poor countries unable to afford the vastly expensive infrastructure compared to one or two satellites. They were also essential to gaining access to the expanding infrastructure of the developed world, giving small countries a chance at accessing markets and money for their goods and service as independent states.

The Shuttle captured much of this market in the first five years of operations, an activity that pushed the pace of Shuttle flight schedules and the demand for more missions. After the loss of Challenger the Shuttle was not marketed for these commercial flights, ordinary expendable rockets launching satellites instead while the Shuttle was reserved for government science satellites, probes to the planets or space station assembly flights. The commercial drive would continue, but using unmanned rockets for launch satellites and for countries to develop their own space programmes.

Partners in space

Development of the Shuttle and the way it was used in space opened NASA to international co-operation it had not known before, allowing other countries to build equipment that would fly on the Shuttle and form a key part of its function. Two of those elements, the remote manipulator arm and the Spacelab laboratory module, were crucial to the success of the Shuttle in many of its operations.

Because NASA needed to move satellites and cargo in and out of the cavernous payload bay, it needed a mechanical arm to grasp equipment and shift it about.

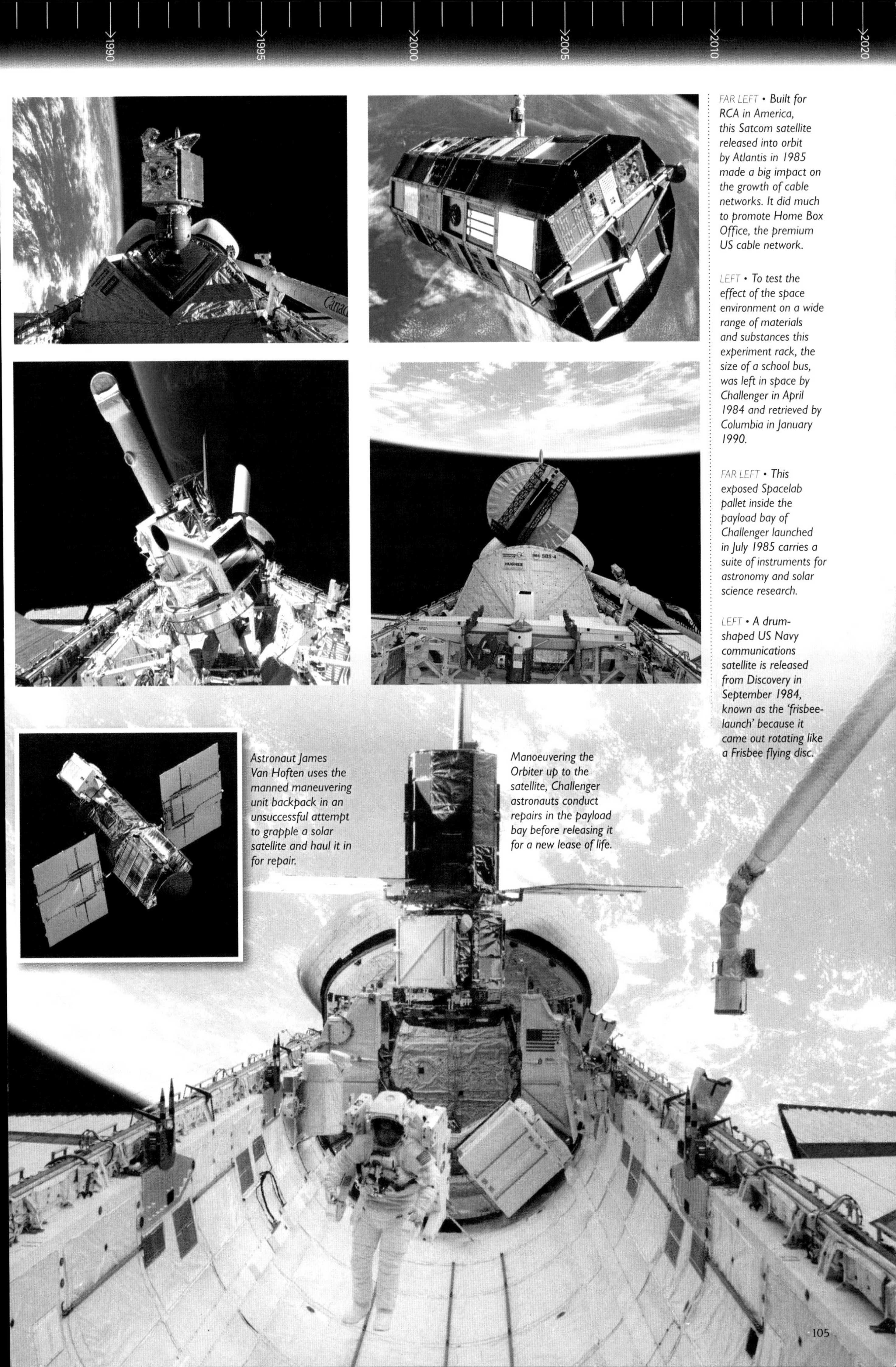

FAR LEFT • Built for RCA in America, this Satcom satellite released into orbit by Atlantis in 1985 made a big impact on the growth of cable networks. It did much to promote Home Box Office, the premium US cable network.

LEFT • To test the effect of the space environment on a wide range of materials and substances this experiment rack, the size of a school bus, was left in space by Challenger in April 1984 and retrieved by Columbia in January 1990.

FAR LEFT • This exposed Spacelab pallet inside the payload bay of Challenger launched in July 1985 carries a suite of instruments for astronomy and solar science research.

LEFT • A drum-shaped US Navy communications satellite is released from Discovery in September 1984, known as the 'frisbee-launch' because it came out rotating like a Frisbee flying disc.

Astronaut James Van Hoften uses the manned maneuvering unit backpack in an unsuccessful attempt to grapple a solar satellite and haul it in for repair.

Manoeuvering the Orbiter up to the satellite, Challenger astronauts conduct repairs in the payload bay before releasing it for a new lease of life.

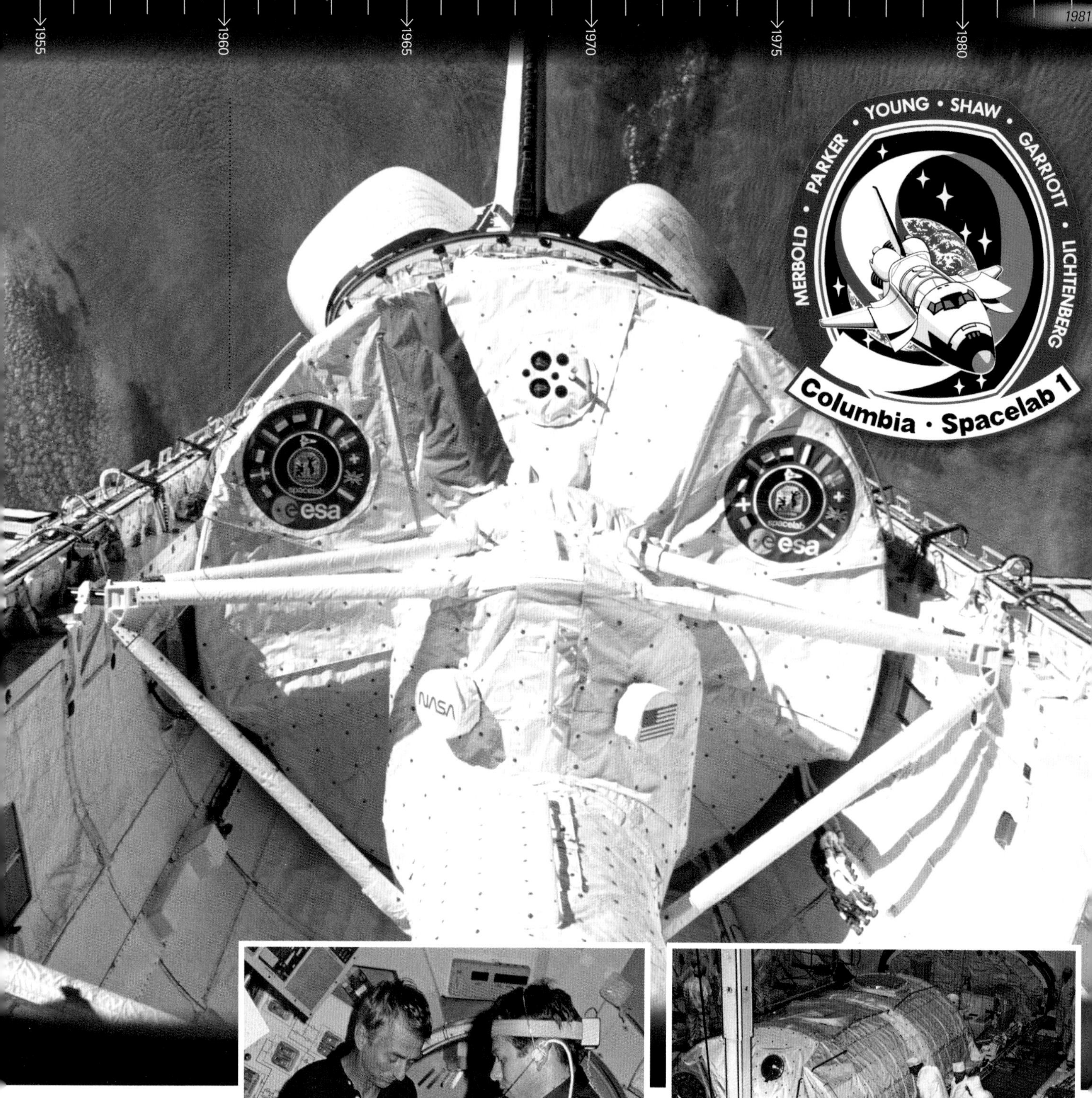

ABOVE • Spacelab inside the cargo bay, seen through the aft facing windows on the Orbiter flight deck.

RIGHT • Aboard Spacelab in November 1983, veteran Skylab astronaut Owen Garriott conducts medical tests on fellow astronauts.

FAR RIGHT • Built by the Europeans, the Spacelab laboratory module is prepared for flight.

Canada's Spar Aerospace, now known as MD Robotics and part of MacDonald Dettwiler, developed and built the manipulator arm under contract to NASA and it has become a crucial part of Shuttle activity on orbit. Consisting of two articulated upper and lower arm sections the remote manipulator is 50ft long and capable of handling massive loads as it maneuvers payloads and freight around in space.

Spacelab consisted of a pressurized laboratory fixed in the Shuttle's payload bay. It was built by the European Space Agency, with most of the work going to Germany. The first Spacelab mission came on the ninth Shuttle flight commanded by veteran astronaut and moonwalker

John Young. Launched in November 1983 the mission lasted just over 10 days, during which the crew carried out extensive scientific tests and experiments.

Spacelab also encompassed pallets, essentially giant instrument trays the full width of the 15ft diameter payload bay carrying experiments designed for exposure to the vacuum of space. In all, the Shuttle flew pressure modules on 16 flights with nine additional flights carrying only pallets. The Spacelab equipment effectively adapted the Shuttle into an interim research laboratory between the last Skylab mission in 1974 and the development of the international space station, assembly of which would begin in 1998, the year the last Spacelab flight was flown.

The Shuttle programme had been a great inspiration for ESA – the European Space Agency – which was set up in the mid-1970s to make European countries independent of the United States in launching satellites and carrying out experiments in space. The Shuttle provided the vehicle within which European industry could develop the equipment that gave it experience in building manned space vehicles of the future. Had it not been for Spacelab and the lesson learned in that work, European industry would not have been able to participate in plans for a giant orbiting space station.

Accompanying the release into orbit of the national satellite Morelos-B, Mexican astronaut Rudolfo Neri-Vela takes a snack in the mid-deck area of Atlantis, November 1985.

The Shuttle space suit is the ultimate development of simple high-altitude pressure suits worn by Mercury astronauts 50 years ago and is capable of supporting space walks lasting up to eight hours.

A bigger place in space

At the annual State of the Union address in January 1984, President Ronald Reagan formally announced NASA's 'next logical step' – to build a permanently manned space station that would be assembled over several years by Shuttle. Later that year, NASA boss James Beggs went on a world tour, inviting other countries to join together in building and working on the space station. By 1985, Europe, Japan and Canada had agreed to join with the United States and work with NASA on this bold new venture. Brazil came in representing a Third World contingent, eager not to miss out on the benefits from this research.

Using its experience working with the Shuttle programme, Canada would build the robotic arms and manipulator systems, Europe would build a laboratory module called Columbus and Japan would also build an experiment module that would be attached to the main structure launched by NASA and assembled in space. By the end of 1985 there were four Shuttle Orbiters in the fleet and a big plan for the world's largest engineering project ever put together involving many countries. Nobody could know that an unforeseen and catastrophic disaster was about to change everything.

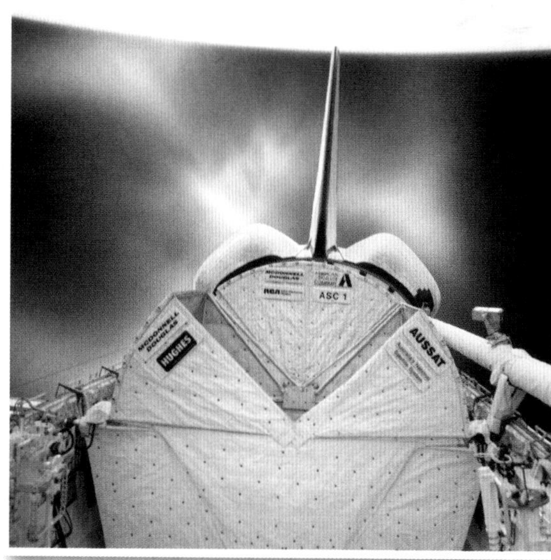

ABOVE • *Looking back across the tail of Discovery as the two orbit manoeuver engines are fired to change trajectory. The nearest payload dock is open, indicating that the Australian satellite it contained has been released, while another for the American Satellite Company has its bay doors shut.*

BELOW • *Simulated in this electron beam arc, the pressure wave on this Shuttle model displays the plasma sheath that builds up during re-entry.*

"BY THE END OF 1985 THERE WERE FOUR SHUTTLE ORBITERS IN THE FLEET AND A BIG PLAN FOR THE WORLD'S LARGEST ENGINEERING PROJECT EVER PUT TOGETHER INVOLVING MANY COUNTRIES."

After landing a fleet of vehicles attend the Shuttle to make it safe for towing back from the runway. Prominent is the pipe boom of the ground coolant vehicle, travelling in convoy as the tractor slowly pulls it away.

TRAGEDY STRIKES

The loss of Challenger and its crew

On January 28, 1986, an unusually cold morning at the Kennedy Space Center, preparations were nearing completion for the 25th Shuttle flight, the tenth flight of Challenger. By noon the Shuttle had been destroyed along with its seven-member crew.

The flight was planned as a science mission to study Halley's Comet, launch a big communications satellite and perform several technology experiments. The crew comprised Michael Smith, Dick Scobee, Ronald McNair, Ellison Onizuka, Gregory Jarvis and Judith Resnik. Also on board was a schoolteacher, Christa McAuliffe. She had been selected to represent teachers everywhere and was preparing to inspire children with televised lessons she planned to conduct from orbit.

The flight had already been delayed several days but when engineers saw icicles and frozen pools of water surrounding the Shuttle they were certain the flight would be cancelled. There was fear that ice would break loose and shatter heat protecting tiles on the Orbiter. However, victims of the pressure to launch as quickly as possible at every opportunity, senior managers decided that the flight would go ahead.

Challenger was launched at 11.38am , January 28, 1986, from pad 39B. At the moment of launch, puffs of smoke from the outer casing on a solid rocket booster revealed a leak of hot gases. At 35 seconds the leak broke into a flame, cutting open the bottom of the external tank like an acetylene torch. At 73 seconds, with

The Challenger flight crew. Back row (left to right), Ellison Onizuka, Christa McCauliffe, Greg Jarvis, Judy Resnik; front row (left to right): Mike Smith, Dick Scobee and Ron McNair

Highly unusual for Florida, even in January, Ice surrounds the gantry at pad 39B on the morning of launch.

Milestones

1986

JANUARY 28:
SHUTTLE CHALLENGER IS DESTROYED 73 SECONDS INTO FLIGHT AFTER LAUNCH ON A PARTICULARLY COLD MORNING, TAKING THE LIVES OF 14 ASTRONAUTS.

FEBRUARY 3:
PRESIDENT REAGAN FORMS A SPECIAL COMMISSION TO INVESTIGATE THE EVENTS.

JUNE 9:
THE ROGERS COMMISSION RELEASES ITS REPORT, CLAIMING COMPLACENCY AND A LACK OF DESIGN IMPROVEMENT AS A FUNDAMENTAL CAUSE OF THE DISASTER.

SEPTEMBER 29:
THE SHUTTLE RETURNS TO FLIGHT WITH THE LAUNCH OF DISCOVERY.

ABOVE • *Teacher Christa McAuliffe was selected from 11,000 applicants to fly in space and inspire children with lessons directly into their class-rooms from orbit.*

RIGHT • *Just barely detectable in the lower right of this frame the black smoke that blew through the booster segment can be seen 03. seconds after ignition.*

BELOW • *Icicles surround the Shuttle on the day of launch. After the disaster a mandatory ice inspection team checks the vehicle and pad area, whatever the weather.*

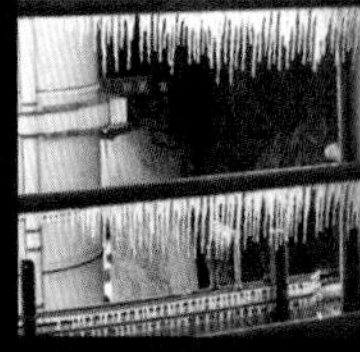

Challenger at an altitude of 48,000ft, the tank was torn to pieces and a violent explosion shoved the two solid boosters to either side.

Challenger began to break up immediately but the momentum kept the debris soaring to a height of 65,000ft, before falling back into the Atlantic Ocean. Shocked, children across the United States watched TV screens in a state of numbed disbelief and President Ronald Reagan cancelled the State of the Union address to make a national broadcast. It would be two years and eight months before a Shuttle flew again.

'An accident rooted in history'

Immediately on seeing the disaster appear on their screens, flight controllers at Mission Control in Houston locked the doors and impounded all the documents, tapes and records of what had happened on that short flight. Once sealed, they evacuated the control centre and turned it over to an investigation immediately put in place by NASA. But the really important evidence would unfold before a special commission headed by former Secretary of State William P Rogers, including eminent specialists such as Neil Armstrong, astronaut Sally Ride, Chuck Yeager and the famous physicist and Nobel laureate Richard Feynman.

What the Rogers Commission heard was a damning indictment of management practices that seemingly ignored warning signs of danger in the continued use of

The Challenger crew on the flight deck at launch with pilots Smith (left) and Scobee. Resnik (right) and Jarvis are behind them.

Above: At 73 seconds into the flight, the two boosters are shoved aside by the force of the eruption as gases and boiling hydrogen fill the region where the Orbiter should be.

flawed solid rocket booster designs. On more than one occasion boosters had shown signs of leaking gases and, in one instance, flame, during flight. Because the boosters were returned to earth and reused, any faults could easily be detected. The initial ignition of Challenger's boosters created a giant shock wave inside the joints between segments. In very cold weather the rubber seals preventing a blow-through would become rigid and brittle.

Richard Feynman demonstrated this when he produced a piece of the same flexible rubber used on the booster joints and dropped it in a glass of ice water. When he pulled it out seconds later it was rigid, quite unable to flex into the joint and seal it. There was clearly a design flaw and the manufacturer was indicted for having ignored this and become complacent. It was, concluded the Rogers Commission, 'an accident rooted in history'.

Christa McAuliffe and (left) her backup Barbara Morgan, who was selected as an astronaut in her own right in 1998 and made it into space aboard Endeavour in August 2007.

The 'Teacher in Space' programme badge from which Christa McAuliffe and Barbara Morgan were selected in 1985.

Return to flight

After months of poring over the evidence and analyzing the data, the Rogers Commission made a set of recommendations that included a robust redesign of the booster segments. A wide range of safety issues were debated and put into practice but it was more than 32 months before NASA was ready to fly again. That event took place in September 1988 when Discovery carried Rick Hauck, Dick Covey, John Lounge, George Nelson and David Hilmers back into orbit.

Much had changed and one of the recommendations of the commission was that NASA should no longer fly commercial satellite payloads due to the pressure to launch on schedule that those imposed for customers around the world. Henceforth, the Shuttle would fly only government payloads for scientific and research purposes but NASA could look toward the day when it could start assembling the space station, which it had now named Freedom. As it turned out that would be a very long wait.

The moment of truth in Mission Control as TV screens show the horror of Challenger's fate.

Substantial parts of the Orbiter were recovered including this power head from one of the three main engines.

The Shuttle returns to space as Discovery lifts off on a successful mission in September 1988.

AN EYE ON THE UNIVERSE

Scientist Lyman Spitzer is considered the 'father' of the Hubble telescope, having fought long and hard to see it through difficult days when Congress tried to cancel it to save money.

Shuttle launches Hubble Space Telescope

The biggest optical telescope launched into space was placed in orbit by Discovery on April 24, 1990. Capable of seeing to the far edge of the Universe and further back in time than any telescope before it, the Hubble Space Telescope had been planned for more than 25 years – and is still providing stunning images of the Universe.

The transparency of earth's atmosphere hides a secret – telescopes on the ground see stars fuzzier than they are in reality because of shimmering effects caused by the air we breathe. Only by placing a telescope in the vacuum of space can scientists see clearly objects that are indistinct when viewed from the earth below. This is why, from the dawn of the space age, astronomers dreamed of sending a powerful telescope into orbit.

Fifty years ago it was thought astronomers would have to be in space too, physically observing stars and galaxies through the telescope's eyepiece as they do on earth. But with the revolution in electronics and digital imagery, a telescope could beam to earth images for scientists to examine without having to go into orbit. So it was that the Hubble Space Telescope was developed to do just that, sending images to the ground obtained from several instruments all looking at objects reflected by a giant mirror on Hubble.

Polishing the mirror. Because of instrument errors during the grinding process it was slightly deformed, ruining images restored to clarity by a subsequent servicing mission.

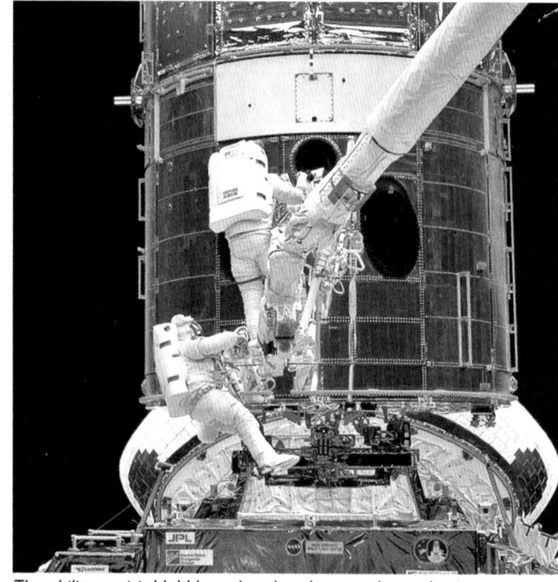

The ability to visit Hubble and replace large and complex instruments has kept the telescope at the forefront of technology, vastly improving its capabilities over the last two decades.

The fuzzy view via the distorted mirror (below) was corrected by several new instruments attached during the first servicing mission, as seen here (bottom).

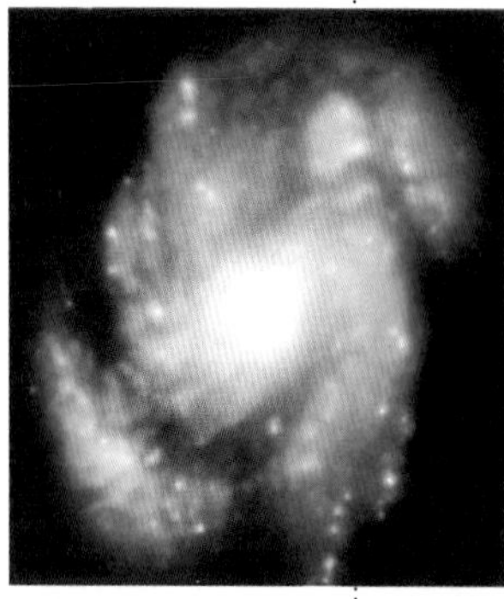

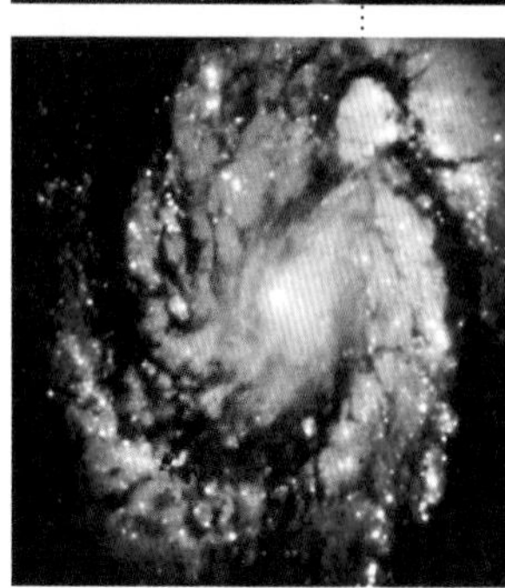

A distorted view of the Universe

The Hubble Space Telescope had been in development since the late 1960s and was scheduled for launch in 1986 when the Challenger disaster grounded all Shuttle Orbiters for more than two years, until finally Discovery was able to lift it into space in April 1990. No sooner had Hubble been launched than it became apparent that there was something wrong. The 92-inch diameter mirror, from which all objects were reflected and collected by the separate instruments, was deformed by a tiny amount at the outer edge, a mere 2.2 thousandths of a millimeter. It was sufficient to completely upset the optical alignments and render the telescope virtually useless.

Mending the mirror

Hubble had been designed for routine access by Shuttle astronauts, servicing and upgrading the telescope as necessary, now there was an urgent need to get back to Hubble and repair the mirror. In December 1993 Endeavour carried into

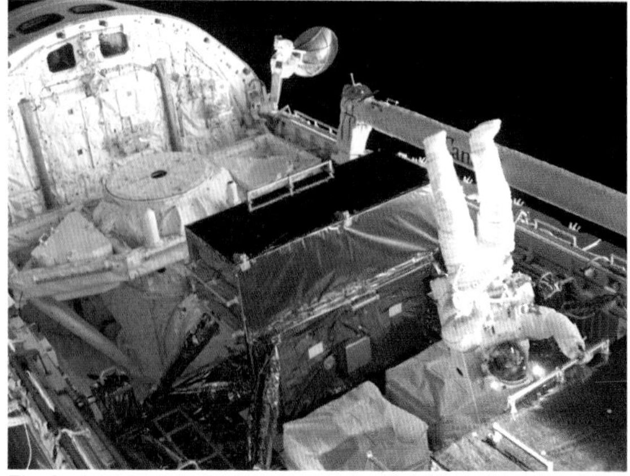

The final servicing mission launched in May 2009 carried a load of spares and replacement equipment to keep Hubble going for at least another five years.

WHAT'S IN A NAME?

Along with Albert Einstein, Edwin P Hubble was one of the most important scientists of the 20th century. Born in the US in 1889, he confirmed the existence of galaxies, established a scale by which the expansion of the Universe could be measured and did more than anyone else to bring astronomy, then considered a separate science, into the realm of physics. Continuously seeking to explain the nature of the Universe itself, Hubble is an appropriate name for the world's first big optical telescope in space.

With a weight of 12 tons, the Hubble Space Telescope is grasped by the Shuttle's 50ft long robotic arm and maneuvered down into the payload bay for servicing.

RIGHT • A Lilliputian view of astronauts dwarfed by the giant telescope. They are working on the final servicing mission to set it up for several more years of useful work.

BELOW LEFT • Hubble is in a nearly circular orbit 350 miles above the earth and takes almost 97 minutes to make one circumnavigation of the globe.

BELOW RIGHT • Hubble has a focal length of 189ft and a 94 inch diameter mirror, its development greatly enhanced as an international venture between NASA and the European Space Agency, which has supplied some of its equipment.

Milestones

1990
APRIL 24: DELAYED BY THE CHALLENGER DISASTER, HUBBLE IS FINALLY LAUNCHED BY DISCOVERY BUT SUFFERS FROM A DISTORTED MIRROR.

1993
DECEMBER 2: ENDEAVOUR FLIES A HUBBLE SERVICING MISSION TO CORRECT THE OPTICS.

1997
FEBRUARY 11 DISCOVERY LAUNCHES A SECOND SERVICING MISSION AND REPLACES EQUIPMENT TO ENHANCE ITS PERFORMANCE.

1999
DECEMBER 19: DISCOVERY FLIES THE THIRD HUBBLE SERVICING FLIGHT.

2002
MARCH 1: COLUMBIA PROVIDES HUBBLE WITH NEW EQUIPMENT AND REPLACEMENT EXPERIMENTS ON THE FOURTH SERVICING MISSION.

2009
MAY 22: ATLANTIS FLIES THE LAST HUBBLE UPGRADE MISSION, PROVIDING A BATTERY OF NEW EQUIPMENT, UPGRADED SENSOR PACKAGES AND REPLACEMENT POINTING DEVICES.

space one of the most highly trained crews ever launched, together with specialist tools to fix Hubble, a corrective optics package named COSTAR. It worked and Hubble's eyes got a 'pair of spectacles' that corrected its vision.

Another servicing mission was launched 1997 when Discovery replaced two instruments with updated versions and old reel-to-reel tape recorders were replaced with digital storage devices. A third visit in December 1999 replaced six ailing gyroscopes that kept it correctly pointed at objects in space. Another visit in 2002 replaced the last remaining original instrument with an improved package, further enhancing its capabilities and giving the telescope a new lease of life.

Final tweaks to a lasting legacy

The last servicing mission took place in 2009 when Atlantis carried a Hubble repair crew armed with two new instruments for enhancing its ability to see objects transmitting in ultraviolet light. The crew also repaired two instruments that had failed, and replaced equipment essential to keeping it accurately pointed at selected objects. More than 13 years beyond their design life, two large batteries were also replaced with new ones.

Hubble had been built with the Shuttle in mind, being fitted with grapple fixtures that would allow the Orbiter's robotic arm to grasp it for servicing in the payload bay. As it turned out that was vital for carrying out repairs and keeping it fitted with the best and most up-to-date instruments available.

Without Shuttle and its heavy lifting ability, plus space-walking crews to service and upgrade it, the world of astronomy would be poorer and the wonder and awe inspired in the general public by its stunning views of the Universe would never have taken place.

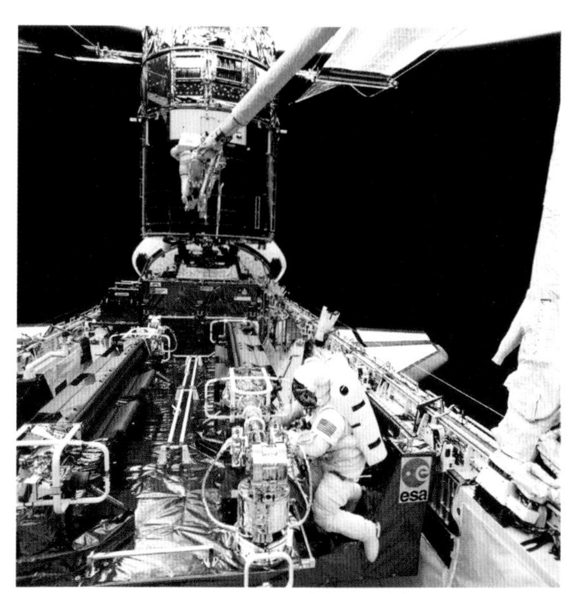

Celebrating two decades of useful astronomy, in April 2010 the Hubble Institute released this view from the telescope of a small part of the Carina Nebula more than 6,500 light years from earth.

Fully assembled, the Mir space station as seen on June 12, 1998 when Discovery departed after the last of nine Shuttle visits.

Selected from 13,000 applicants, on May 18, 1991 Dr Helen Sharman became the first British citizen flown into space when she was launched by Soyuz for a week-long stay aboard Mir.

MIR

RUSSIA's HOME IN SPACE

It took years to develop but when assembled in orbit it was the biggest workplace in space.

It came after two decades of experimenting with smaller research laboratories, but when Russian space scientists launched the core module for their Mir space station on February 19, 1986, it was a whole new ball game. Over a period of 10 years the Russians built Mir into the world's largest space laboratory and set records for space flight duration that stand today as the longest in the history of human space travel.

When complete with all its separately launched modules attached, the Mir complex weighed almost 130 tons and had a length of 62ft, a width of 102ft and a height of more than 90ft. It had a pressurized atmosphere similar to that at the earth's surface and orbited the earth at a height of more than 200 miles.

Staying in space

Since the end of the Apollo moon race, Russian space scientists searched for ways to study the effects on the human body of very long periods of weightlessness. So long, that it would equal the cumulative 18 months cosmonauts would spend on a trip to Mars and back. Mir became the perfect place to test human reactions to long flights. It was known that after flights of only a few weeks there was a marked deterioration in bone density, blood cell levels and in the ability of muscles to remain toned and effective. Exercise regimes were vital for maintaining the ability of cosmonauts to return to earth in a healthy condition.

What the Russians conducted was a test in human endurance and in doing so they made many contributions to medical science that helps reduce the effects of ageing in people on earth. In effect, when people go into space they experience the effects we all encounter as we get old and suffer the consequences of brittle bones, reduced immunity to disease and longer periods needed to heal injuries. To gather this information, Russian scientists sent cosmonauts into space for increasingly longer flights. Launched on January 8, 1994, Valeri Polyakov added to a previous Mir flight of 240 days by remaining aboard the Russian space station for a full 437 days, bringing to a total of 677 days his collective experience of weightlessness.

A special badge to mark the eleven docking flights between Shuttle and Mir.

Milestones

1986
FEBRUARY 19: RUSSIA'S MIR CORE MODULE IS LAUNCHED BY PROTON ROCKET.

1987
MARCH 31: THE FIRST OF SIX ADDITIONAL MIR MODULES IS LAUNCHED TO THE SPACE STATION.

1995
JUNE 29: ATLANTIS DOCKS WITH MIR, THE FIRST US-RUSSIAN LINKUP SINCE APOLLO-SOYUZ IN 1975.

1996
MARCH 22: ATLANTIS IS LAUNCHED TO MIR, BEGINNING A CONTINUOUS US PRESENCE ABOARD THE SPACE STATION UNTIL JUNE 12, 1998.

1998
JUNE 4: DISCOVERY CONDUCTS THE NINTH AND LAST SHUTTLE DOCKING TO THE MIR SPACE STATION.

2000
APRIL 4: THE RUSSIANS SEND THEIR LAST CREW TO MIR, LANDING JUNE 16.

2001
MARCH 23: MIR WAS DELIBERATELY DE-ORBITED TO A FIERY RE-ENTRY AND DESTRUCTION.

RIGHT • The Mir complex carried several solar cell wings with those on the core module providing up to 9kW of power, others on the experiment modules supplementing this supply

BELOW • The Mir badge worn by its cosmonauts.

Shuttling to Mir

Beginning in 1988, the Russians began flying west European astronauts to Mir, starting with France's Jean-Loup Chretien and then Helen Sharman, Britain's first astronaut, in 1991. Astronauts from Austria, Germany and Slovakia followed and in 1993, after the collapse of the Soviet Union, President Bill Clinton offered a hand of friendship to the Russians by suggesting that they join with America's big space station project. This was initiated by President Ronald Reagan in 1984 and named Freedom, but now to be known as the International Space Station. To prepare the way for such close co-operation there would be several Shuttle visits to Mir.

In 1994 the first Soviet cosmonaut to ride aboard the Shuttle, Sergei Krikalev, had been carried to orbit by Discovery. NASA missions to Mir began in 1995 with the launch of Discovery for a fly-around, without docking, to check out rendezvous techniques, communications between the station and the Shuttle, and coordinated mission operations between Houston and Moscow. The first docking with Mir occurred in June 1995, delivering cosmonauts Solovyev and Budarin to the space station, the first time in 20 years that US and Russian spacecraft had docked together. By June 1998 when the last Shuttle flight to Mir took place US astronauts had visited the Russian station nine times, often carrying cosmonauts up, retrieving some already at the station, or leaving US astronauts in their place.

RIGHT • A last look back as Shuttle Discovery separates at the end of three years of joint operations by US astronauts and Russian cosmonauts, June 1998.

Time to go

By the time Shuttle flights to Mir ended in June 1998 the space station was getting well worn and long past its use-by date. It had suffered fires, collision from unmanned Progress logistics freighters, systems breakdown and power failures. But this was the longest lived manned space station of all time and by the time it decayed back down into the earth's atmosphere and broke up, on March 23, 2001, it had been in space for 5,519 days, a period in which it had been occupied for 4,592 days – more than 12 years! In that time, 39 manned spacecraft, including NASA's Shuttle had docked to the station while a further 72 unmanned modules and

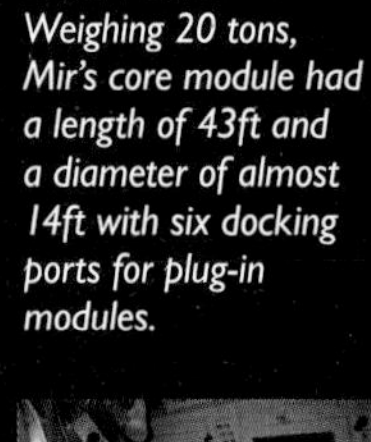

BELOW RIGHT • Weighing 20 tons, Mir's core module had a length of 43ft and a diameter of almost 14ft with six docking ports for plug-in modules.

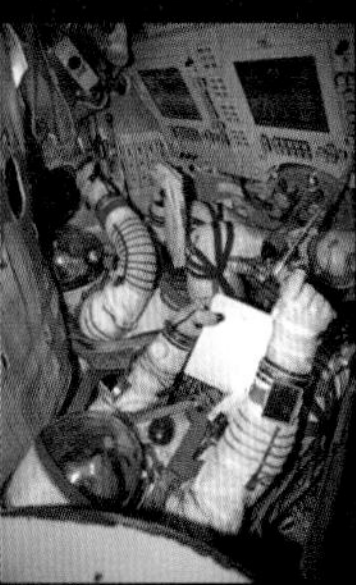

ABOVE • An extendible stick brings switches and control panels within reach of the cosmonauts restrained in their couches on Soyuz.

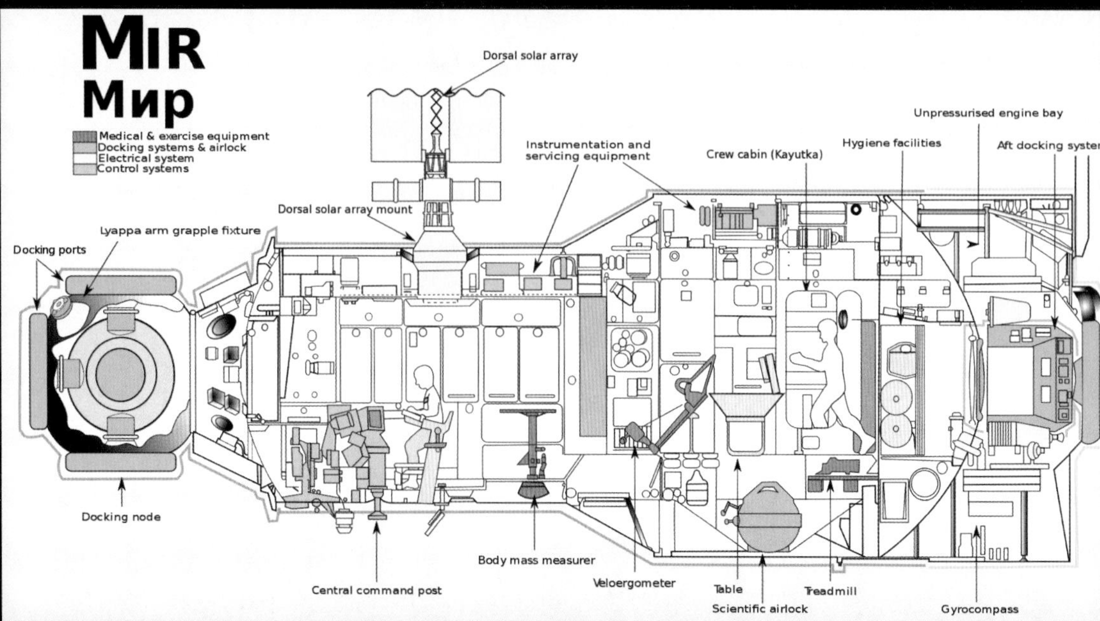

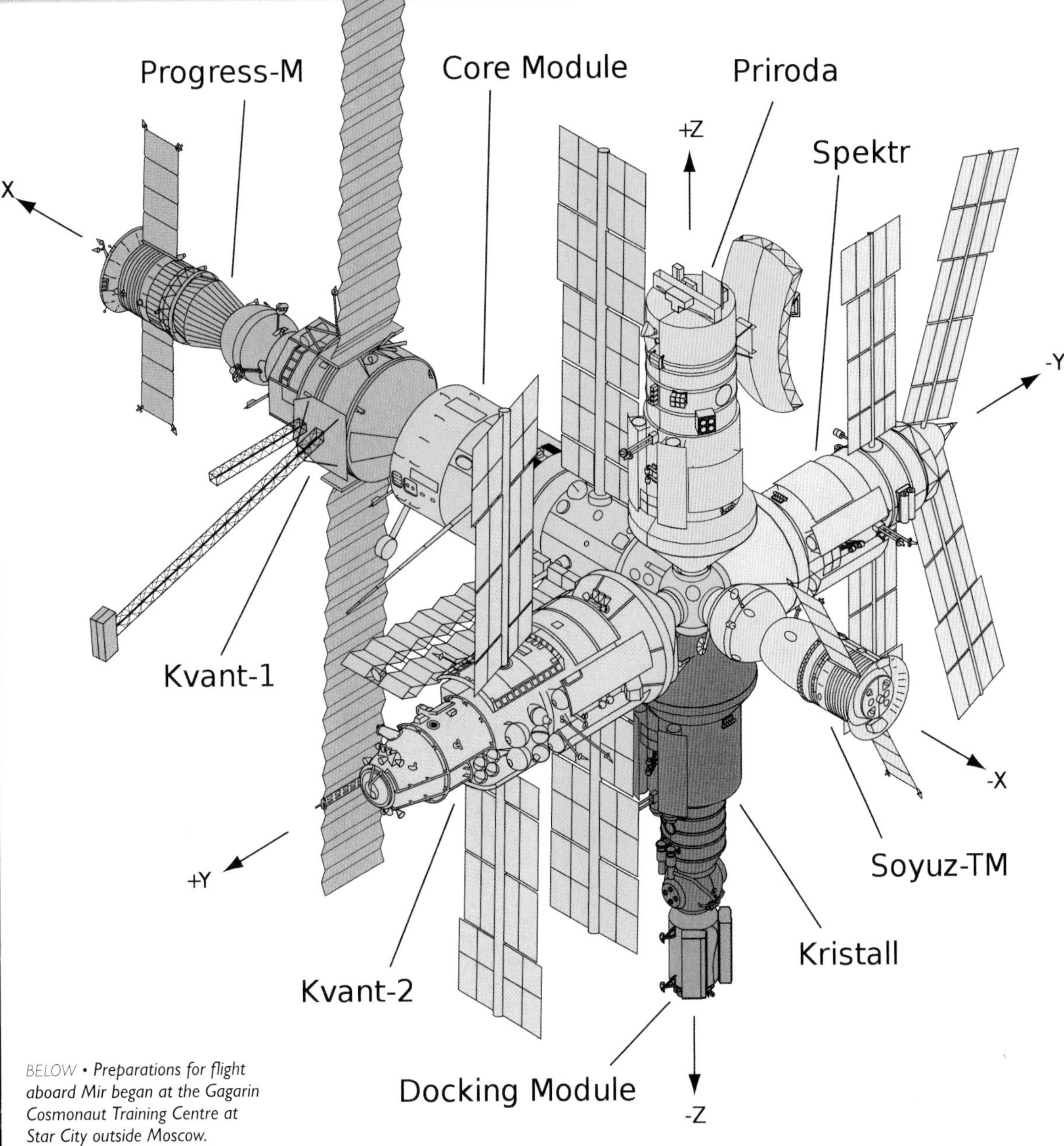

LEFT • The Mir core module served as the main living quarters and operations centre with additional scientific modules attached over several years. Clockwise, Priroda contained earth environmental monitoring equipment, Spektr was launched to suppport US visits and contained living quarters. Kristall had small furnaces and equipment for making new materials. Kvant-2 carried a wide variety of science equipment and Kvant-I supported astrophysics experiments and telescopes. At the front, Soyuz carried crew up and down to Mir, at the rear Progress contained cargo.

BELOW • Preparations for flight aboard Mir began at the Gagarin Cosmonaut Training Centre at Star City outside Moscow.

BELOW MIDDLE • Designed for zero-gravity, the Russian space toilet has a vacuum suction for removing waste products to a sealed container.

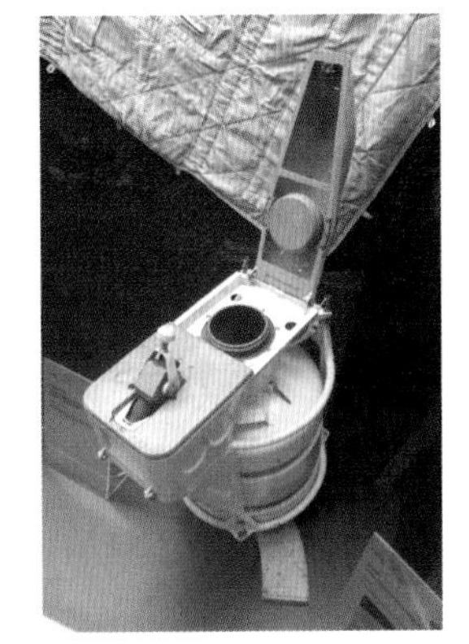

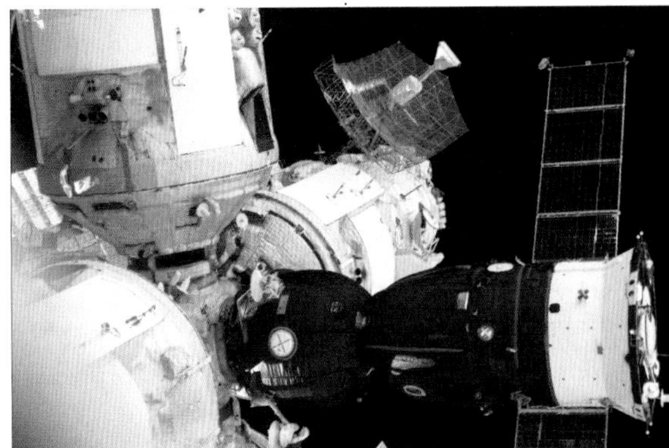

ABOVE: • A Russian Soyuz remains docked to Mir while the crew works in the space station.

LEFT • NASA astronaut Shannon Lucid works out on the Mir treadmill to mitigate the effects of weightlessness.

ad carried additional living space or supplies to Mir, a grand total of 111 spacecraft.

Now it was time for Russia and America to put their space expertise to global use, bringing alive the International Space Station that NASA, Europe, Japan and Canada had been pledged to build since 1984. Out of the political weapons of the Cold War had been forged the ploughshares of a new age of co-operation and discovery, linking old adversaries in a new adventure bringing unprecedented benefits to people on earth and providing new horizons to explore. A preparation for the next logical step – the human exploration of Mars.

INTERNATIONAL SPACE STATION

In a joint endeavour involving 15 countries, former adversaries build a research station for the future – 250 miles above the earth.

Milestones

1998

NOVEMBER 20: RUSSIA'S ZARYA IS LAUNCHED INTO ORBIT AS THE FIRST STATION ELEMENT.

DECEMBER 4: NASA'S UNITY MODULE IS LAUNCHED FOR DOCKING TO ZARYA.

2000

JULY 12: A PROTON LAUNCHES ZVEZDA, RUSSIA'S SECOND MODULE.

2001

FEBRUARY 7: THE PRIME US RESEARCH LAB, DESTINY, IS LAUNCHED.

JULY 12: ATLANTIS LAUNCHES THE QUEST AIRLOCK MODULE.

SEPTEMBER 14: THE RUSSIAN PIRS DOCKING MODULE IS LAUNCHED.

2007

OCTOBER 23: THE SECOND NODE, HARMONY IS LAUNCHED BY DISCOVERY.

2008

FEBRUARY 7: EUROPE'S COLUMBUS LAB LAUNCHED ABOARD DISCOVERY.

MARCH 11: JAPAN'S KIBO PRESSURIZED LAB CARRIED TO THE ISS BY ATLANTIS.

2009

NOVEMBER 10: RUSSIAN POISK DOCKING AND AIRLOCK MODULE LAUNCHED.

2010

FEBRUARY 8: US NODE, TRANQUILITY AND A CUPOLA WINDOW SECTION LIFTED BY ENDEAVOUR.

MAY 14: ATLANTIS CARRIES THE RUSSIAN RASSVET RESEARCH MODULE TO THE ISS.

2011

FEBRUARY 24: EUROPE'S LEONARDO MODULE LIFTED BY SHUTTLE DISCOVERY.

MAY 15: ALPHA MAGNETIC SPECTROMETER CARRIED UP BY SHUTTLE ENDEAVOUR.

2012

MAY 22: FIRST SPACEX DRAGON CARGO CAPSULE DOCKS TO ISS.

2013

SEPTEMBER 18: FIRST CYGNUS CARGO MODULE DOCKS TO ISS.

2016

APRIL 8: BIGELOW EXPANDABLE ACTIVITY MODULE CARRIED UP BY SPACEX FALCON 9.

2020

MAY 30: FIRST CREW DRAGON TO CARRY ASTRONAUTS TO THE ISS.

Above • The ISS showing the main truss assembly. In front right, the Japanese laboratory modules and at left the European Columbus module. In front are the US Harmony and Destiny modules.

From the time humans first went into space, the impact on the body was worrying because studies indicated that there could be irreversible effects. Loss of calcium made bones brittle and changes to the blood plasma made the body less immune to disease and illness. Over-production of water and weakened muscle fibre threatened healthy hearts and body stamina and weakened limbs made it difficult to return to earth's gravity.

Other changes were more intriguing and gave the potential for benefits. In the 1970s NASA's Skylab and later Europe's Spacelab, launched inside the Shuttle on flights during the 1980s and 1990s, gave scientists opportunities to test these effects on materials and medicines. Coming together on a space station, astronauts could research new levels of science and technology for applications on the ground. With a global reach, the International Space Station would allow experiments developed on earth to be tested and analysed during or after their operation on the ISS.

And then there was the expanding base of technical knowledge, practicing assembling giant structures in space, performing extended periods of work outside the station, developing ways of keeping people alive and healthy over very long periods. There was the possibility that the International Space Station (ISS) could be a stepping stone back to the Moon and on to Mars.

Sizing it up – building it big

When first announced by President Reagan in 1984, the station was to be assembled by the US and a collection of friendly countries. Europe, Japan, and Canada eventually signed up to help build the station, forming what would be known as the US Orbital Segment (USOS). Russia joined in 1994 after the collapse of the Soviet Union with its section known as the Russian Orbital Segment (ROS).

ABOVE • *Originally planned as the core for Mir-2, the Zvezda module was rebuilt for the Americans and launched in 2000.*

ABOVE • *Russia's mission control facility for the ISS as seen on April 12, 2007 during ISS operations.* BELOW • *Astronaut Williams in a crowded Destiny lab on the ISS, packed with scientific equipment.*

NASA would rely on the heavy lifting capacity of the Shuttle to put up US, European and Japanese pressure modules while the Russians would use the giant Proton rocket to send up their own modules separately. Europe would use its Ariane launch vehicle to lift cargo modules designed to rendezvous and dock with the station and NASA would launch logistics canisters built by Italy. Japan would use its own H-IIB rocket to lift dedicated cargo pods to its experiment module attached to the ISS.

Assembly began with the launch of the Russian Zarya module in November 1998. Built by Boeing, NASA's Unity docking module was launched by Endeavour the following month and this was followed by another Russian module, Zvezda, in July 2000. On its side was an advertisement for Pizza Hut – for which they allegedly paid $1m! Zvezda had been built as the core module for Mir-2, but that project was abandoned at the collapse of the Soviet Union and made redundant when Russia joined the ISS.

The first expedition to the ISS arrived on 2 November 2000 and it has been occupied ever since, hosting a total of more than 250 people from 19 different nations to date. Most crew members remain aboard the ISS for approximately six months and over the time the number of astronauts on board has fluctuated between two and six, with seven from 2020.

One by one the various pressurised modules and solar array panels were launched to the ISS. NASA's Destiny module was lifted to the ISS by Atlantis in 2001, followed by the second core module, Harmony launched by Discovery in 2007. Europe's Columbus laboratory was attached to the ISS after launch by Atlantis in 2008 followed by Japan's Kibo module later the same year. Early in 2010 Tranquility, the third and final node, was lifted aboard the station by Endeavour. Built in Italy it was handed over to NASA as a permanent fixture at the ISS later that year. One of the European cargo modules, Leonardo was added in 2011 followed by the last module, the expandable Bigelow module in 2016.

Housekeeping

Keeping the ISS up and running demands a lot of cargo carried to the station in modules looking like giant cylindrical dustbins. Built in Italy on behalf of the European Space Agency, and in Japan as part of its contribution to the space station, a flotilla of resupply ships were launched by these independent states. Added to which are a routine flow of Russian Progress cargo tankers providing food, water, laundry, general provisions, and a variety of scientific experiments.

"ON ITS SIDE WAS AN ADVERTISEMENT FOR PIZZA HUT – FOR WHICH THEY ALLEGEDLY PAID $1M!"

Veteran of a Shuttle and a Soyuz flight, chemist and astronaut Tracy Caldwell Dyson finds the room with the view to do some earth-gazing.

MAIN • Don't look down! Some 200 miles above the Cook Straits, astronaut Curbeam is literally suspended in space as he positions himself at the end of one of the main truss assemblies.

ABOVE • From the flight deck of Atlantis, astronaut Tony Anotelli gazes at the ISS ahead.

TOP LEFT • British born astronaut Michael Foale, an astrophysicist from Louth in Lincolnshire with dual citizenship, has flown six Shuttle missions, a Soyuz spacecraft and conducted extended stays on both Mir and the ISS.

LEFT • Dwarfed by the Canadian built mobile robotics base assembly, an astronaut works his way along the truss assembly to a new work station.

LEFT • Just below a docking probe and target disc astronaut Cassidy performs an EVA during station assembly.

BELOW LEFT • Japan's Kibo module with robotic arm to attach and remove experiments left outside on an exposed pallet.

BOTTOM LEFT • Chow time as nine spacemen – plus the cameraman – gathers in the ISS for a meal where events of the day can be discussed and schedules for the next day reviewed.

BELOW • Play time! Literally as astronaut Lu hammers out a doubtful melody on the keyboard while Duque adjusts the music, relaxing at the end of a work day in the Destiny lab.

BELOW RIGHT • Phillips attends to the Elektra device for converting water into oxygen for air to breathe, a vital chore aboard the space station.

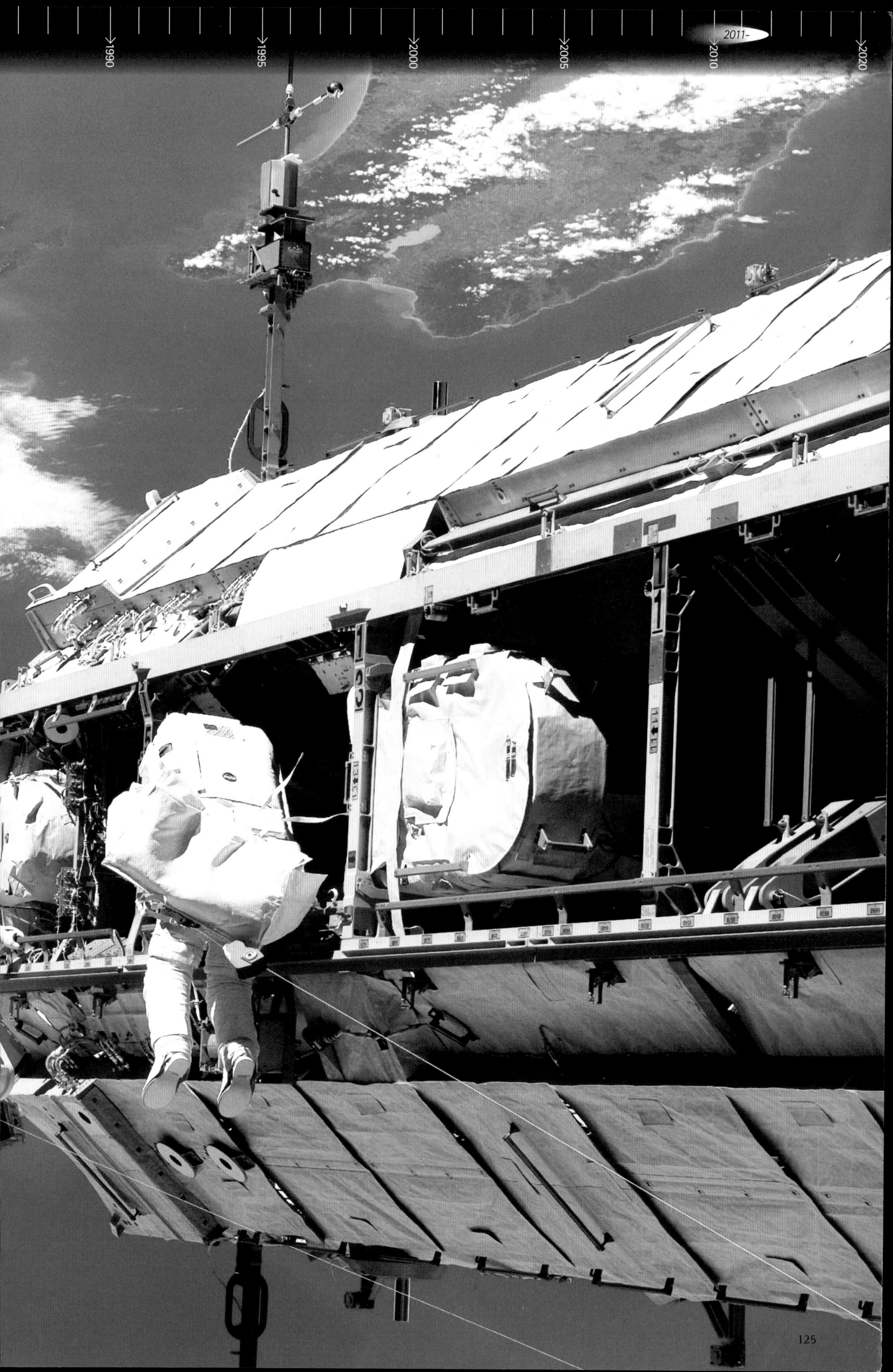

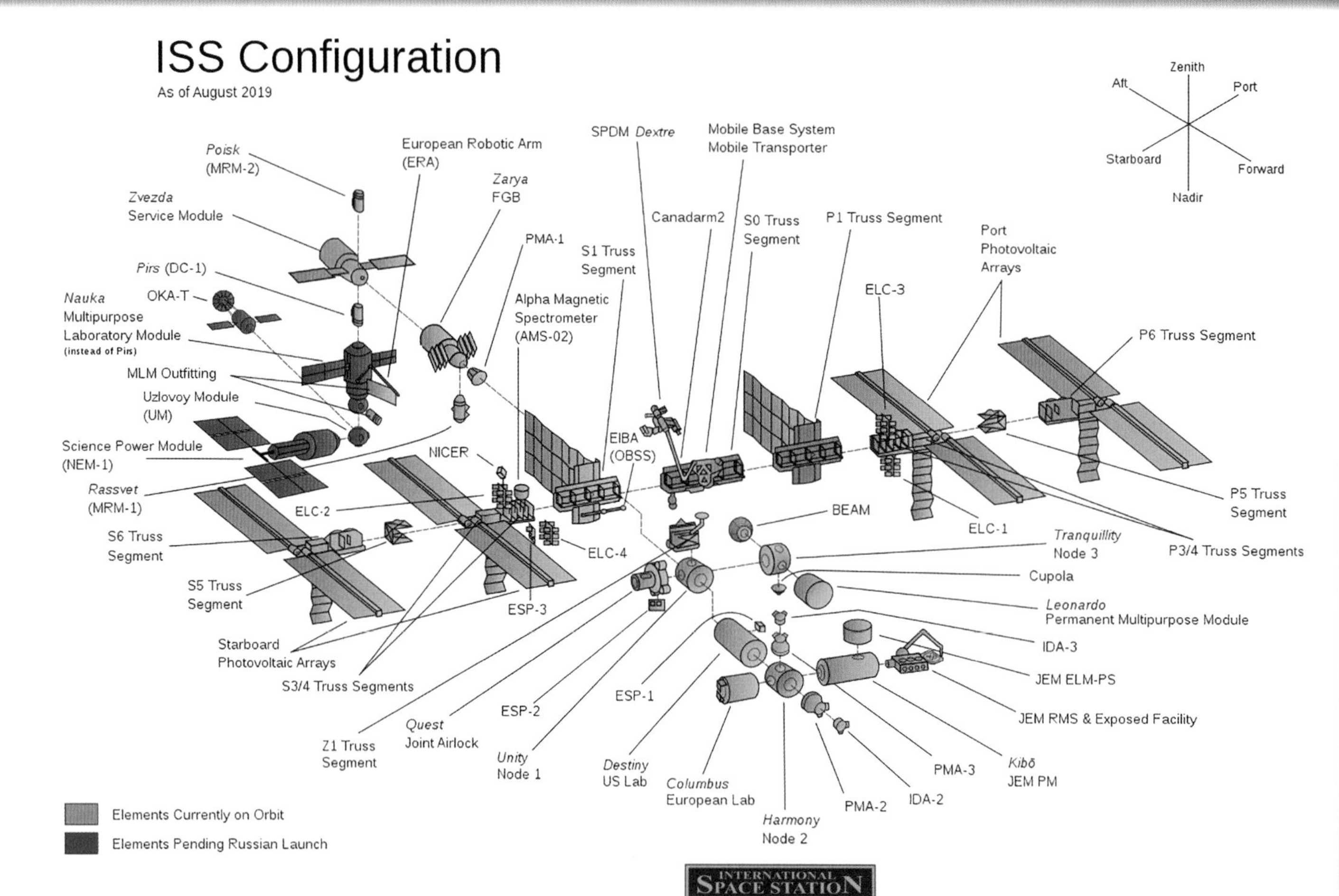

ABOVE • The ISS is the biggest structure ever assembled in space. Bigger than an American football field it has a length of 167ft, a width of 357ft and a height of 66ft with a mass of 400 tons.

In addition, until 2011 astronauts and cosmonauts were carried to the ISS by the NASA Shuttle and by Russia's Soyuz spacecraft. When the Shuttle retired in 2011, only Russia had the ability to carry people to and from the orbiting laboratory – until commercial provider SpaceX began to fly cargo in its Dragon capsule from May 2012 and astronauts in Crew Dragon capsules from May 2020. From 2021 non-Russian astronauts will fly exclusively in Crew Dragon, with Boeing's Starliner capsule planned to carry crewmembers from 2022.

Operating the ISS requires a lot of electrical power and that is provided by batteries charged through the solar arrays, eight giant 'wings' each 112ft long and 39ft wide mounted in pairs with four paired arrays at each end of the truss assembly. Each wing carries 32,800 solar cells and is capable of producing 32.8kW of electrical power.

The truss is the backbone to which all elements are attached and comprises 11 sections of varying lengths locked together end-to-end. The rotating solar arrays are attached to opposite ends of the truss structure and move to track the sun but 35 minutes of each 90-minute orbit is spent in darkness, so the nickel-hydrogen batteries ensured a continuous supply of electrical energy. In 2016 they were replaced by lithium-ion batteries.

To provide the crew with an earth like atmosphere of oxygen and nitrogen, the ISS has adopted innovative technologies. For instance, to provide essential oxygen, a special device uses electrolysis to separate oxygen and hydrogen from used water, recirculating the oxygen into the pressure modules occupied by the crew and dumping the hydrogen overboard.

Built to last – but for how long?

It took many years to build the ISS and the partners intend to operate it until at least the year 2030, probably longer. As the ISS frequently sinks back down toward the atmosphere it is periodically boosted into a higher orbit using the propulsion system on the Russian Progress cargo vehicles or thrusters on the Zvezda module. Its life is limited only by the survivability of its equipment and essential systems on board.

Major strides have been taken by commercial companies in the US to supply the ISS with logistics and send crew to the station. The biggest contribution comes from Elon Musk's SpaceX company which has developed a series of Falcon rockets and Dragon capsules. Supported by some government money to develop these rockets and spacecraft, and pay for its services, SpaceX is now one of the world's major launch vehicle providers, with NASA being a customer rather than an owner. All of which leaves the space agency to concentrate on exploration

BELOW • Endeavour hooks up to the ISS amid robotic arms essential to moving cargo and logistics modules.

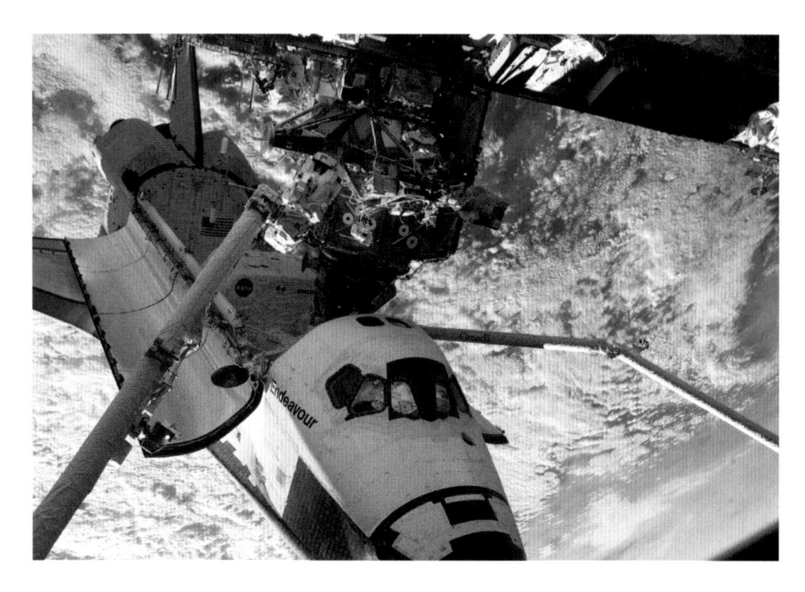

and deep space flight. But there are others, including Northrop Grumman with its Cygnus cargo module and Boeing now developing its Starliner crew-carrier

For the time being, the ISS is the world's premier human space flight programme, which has supported research that has greatly benefitted humans on earth. What happens to astronauts on long flights in weightlessness, mimics the physical changes as people on earth get older. NASA now has one of the largest biophysical databases on the planet, feeding information to medical facilities around the world, improving health care for the aged and helping fight disease through research in space.

For more than 20 years the ISS has been permanently occupied by men and women from many different countries, demonstrating a collective support for a common purpose, crossing political and ethnic boundaries to work together. The ISS may be the springboard from which further cooperation flourishes as the partner countries unite again in a successor to the space station, one which could carry astronauts to the moon and beyond. Part of that future is to build a small station which orbits the moon, giving astronauts the opportunity to conduct research around another world in space and to directly support operations down to and on the surface.

THE COMEBACK KID!

On October 29, 1998, the first American to orbit the earth returned to space aboard the shuttle Discovery. John Glenn was 40 years old when he made his Mercury flight on February 20, 1962. Called Friendship 7, his spacecraft completed three orbits of the earth during a flight lasting just under five hours. He was 77 when he returned to space, and loved every minute of it. His flight aboard Discovery lasted more than eight days and completed 134 orbits. Between missions he spent a distinguished career in politics, being elected a Democratic Senator from 1974 to 1999 and receiving the respect of his peers in both parties.

When he made his flight, John Glenn was, and remains, the oldest man to have flown in space.

Just 77 years of age, John Glenn makes a return to space in October 1998 more than 37 years after making the first US orbital flight in Friendship 7.

ABOVE LEFT • *The ATV uses its rocket motors to push the ISS to higher orbit, as well as carrying logistical supplies.*

LEFT • *From the rear of the ISS are the two large Russian modules Zvezda and Zarya (right) to which is docked a Soyuz spacecraft. Below, a Progress cargo tanker and another Soyuz can be seen.*

BELOW LEFT • *Cosmonauts Korzun (left) and Treschev unpack the treadmill device essential for the daily workout.*

COLUMBIA DOWN!

Assembly of the International Space Station was slowed by the loss of Columbia on February 1, 2003. It was returning home from a 16-day mission during which the crew had been conducting independent science experiments not connected with the space station. The fate of Columbia had been sealed during launch when some insulation fell off the external tank and struck the fragile carbon-carbon wing leading edge, cracking open a gaping hole. During re-entry, this hole allowed heat into the aluminium structure of the Shuttle's wing and the Orbiter burned up in the atmosphere. The Shuttle returned to flight with the launch of Discovery on July 26, 2005.

The crew of Columbia on its 28th and final mission. Rear (left to right) • David Brown, Laurel Clark, Michael Anderson and Ilian Ramon. Front (left to right) • Rick Husband, Kalpana Chawla and William McCool.

TO THE MOON AND BEYOND

It has taken only 60 years to go from a one-man capsule on a single orbit of the earth to a 400 ton space station built to last 30 years. What's next?

NASA's Orion spacecraft consists of a crew module occupied by the astronauts and a service module shown here in its initial configuration with disc-like solar cell panels. (NASA)

The final configuration of the Orion spacecraft envisaged heading for the moon. (NASA)

After 135 launches, NASA retired the Shuttle in 2011, five years after it began sponsoring private companies to come up with cheaper ways to fly to and from orbit. The Russians continued to fly people to the International Space Station in Soyuz spacecraft owned and operated by the government. In the United States, a number of commercial companies received seed money, added to their own financial investment, to produce rockets and spacecraft capable of carrying people into space on a contractual basis, which they began in 2020. Instead of owning the hardware, NASA would hire services from those companies.

Leaving the commercial contractors to operate services back and forth to the ISS in low earth orbit, NASA began development of a spacecraft called Orion capable of carrying astronauts to the Moon and beyond. Plans on how to use Orion changed over time, first being assigned to a Moon landing programme, then to a programme aimed at retrieving asteroid samples, finally back to supporting a return to the lunar surface for the first time since 1972.

Pushing the boundary

In recent years, NASA has decided that a return to the moon will allow international partners to build a range of spacecraft with which astronauts can develop skills for moving farther into the solar system, including Mars. The same private/public sharing that supports the ISS will be used in assembling a mini-space station in moon orbit called Gateway, visited by the Orion spacecraft launched on a super-booster known as the Space Launch System (SLS). More powerful than the Saturn V which sent Apollo astronauts to the moon, the SLS is scheduled to make its first unmanned test flight by early 2022, perhaps earlier.

A significant shift is made with Orion – America's flagship spacecraft – in that it is a cooperative effort between NASA and the European Space Agency. Europe is building the service module which sits under the crew compartment, built by Lockheed Martin, and provides all the environmental equipment, power supply, rocket motors and communications systems. Orion will carry crew to and from lunar orbit or the Gateway, but a commercially developed moon lander will take astronauts down to the surface.

The plan under what has been named the Artemis programme, is for an initial return landing on the moon to be made within the next several years, followed by a sustainable base on the surface, or at several locations, to survey the sites and conduct initial tests with mineral extraction – a mining operation that could be well under way within the next decade or so. Meanwhile, the Gateway will be assembled and serve as an assembly point for spacecraft, rockets and landers, elements of which will be developed for the next leg of the journey – out to Mars.

ABOVE • NASA astronauts Bob Behnken and Doug Hurley made the first commercial flight to the space station on 30 May 2020. (NASA)

The big journey

After the moon landing in 1969, most people believed humans would be on Mars by the end of that century. But that was not to be. The big budgets and the political will that had driven superpowers to head for space evaporated, replaced by a fixed budget that dwarfs the money spent to get to the lunar surface in the 1960s. Where once it spent more than 2.5% of the national budget each year, for many years now NASA has received only 0.5% per annum. But the dream is still there.

SpaceX owner Elon Musk is building a truly giant rocket named Starship, bigger and more powerful than NASA's SLS, which could be capable of sending astronauts to Mars and beyond. NASA too has Mars in its sights as the next big goal for the coming decades.

But there are limits to how far humans can travel. With our present rockets it would take a spacecraft from earth almost 10,000 years to reach the nearest star, Alpha Centauri. There could be ways to speed that up, but not by much.

For at least the next century, it is likely that space flight will become increasingly routine and within reach of most people. The new technologies now being developed may bring true aerospace vehicles capable of linking places on extreme sides of the earth within one hour travel time and flights into space on a timetable that today appear impossibly frequent.

One thing is certain, for humans the journey has only just begun.

BELOW • Not too far in the future, giant spaceships like Starship could routinely ply between earth and destinations in deep space. (SpaceX)

Windows on the past... ...views of the future

For additional information and for useful places to see more of spaceflight history the following organizations and websites will be of help.

The British Interplanetary Society
27/29 South Lambeth Road
London SW8 1SZ
Tel 020 7735 3160
www.bis-spaceflight.com
Formed in 1933, the Society is open to all and has two publications available on subscription. Holds regular meetings with lectures and has a library open to members.

National Space Centre
Exploration Drive
Leicester LE4 5NS UK
Tel 0845 605 2001
www.spacecenter.co.uk
An educational experience covering the technology of space flight together with descriptions of missions and vehicles.

The Science Museum
Exhibition Rd
South Kensington SW7 2DD UK
Tel 0870 870 4868
www.sciencemuseum.org.uk
The space gallery here has a wide range of exhibits including models and artifacts as well as descriptive displays and a full scale mock-up of a Lunar Module.

Kennedy Space Center Visitor Complex
SR405 Kennedy Space Center
Florida 32899 USA
Tel 001 866 737 5235
www.kennedyspacecenter.com
Situated close to the Kennedy Space Center comprises a large park with exhibits, displays, rockets, Shuttle mock-up to enter, exhibits, and guided tours.

National Air and Space Museum
6th and Independence Avenue SW
Washington DC 20560 USA
Tel 001 202 633 1000
www.nasm.si.edu
Packed with aeronautical and space related exhibits and displays.

Steven F Udvar-Hazy Center
14390 Air and Space Museum Parkway
Chantilly, VA 20151 USA
Tel 001 202 633 1000
www.nasm.si.edu/udvarhazy
Has Shuttle Orbiter OV-101 Enterprise on display.

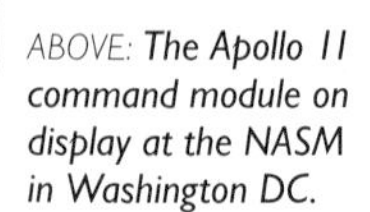

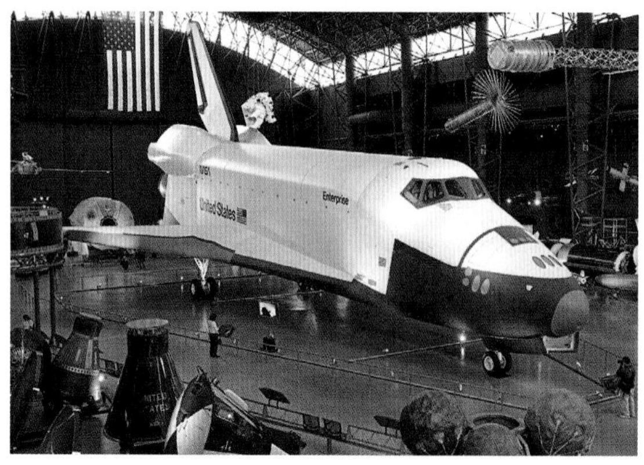

ABOVE: The Apollo 11 command module on display at the NASM in Washington DC.

RIGHT: The Udvar-Hazy Centre at Washington Dulles International Airport, Virginia has the first Shuttle Orbiter, Enterprise, on display.

TOP: The back-up Skylab on display at the NASM.

FAR RIGHT: The X-15 played a vital role in pioneering a path to space.

Websites

WWW.NASASPACEFLIGHT.COM
WWW.SPACEFLIGHTNOW.COM
WWW.NASA.GOV
WWW.ESA.INT
WWW.FEDERALSPACE.RU
WWW.JAXA.JP
WWW.ASC-CSA.GC/ENG
WWW.UKSPACEAGENCY.BIS.GOV.UK